THE ANALYSIS OF NATURAL WATERS

Volume 1

Complex-Formation Preconcentration Techniques

The Analysis of Natural Waters

Volume 1

Complex-Formation Preconcentration Techniques

T. R. CROMPTON

Oxford New York Tokyo
OXFORD UNIVERSITY PRESS
1993

Oxford University Press, Walton Street, Oxford OX2 6DP
Oxford New York Toronto
Delhi Bombay Calcutta Madras Karachi
Kuala Lumpur Singapore Hong Kong Tokyo
Nairobi Dar es Salaam Cape Town
Melbourne Auckland Marid
and associated companies in
Berlin Ibadan

Oxford is a trade mark of Oxford University Press

Published in the United States
by Oxford University Press Inc., New York

A catalogue record for this book is available from the British Library

Library of Congress Cataloging in Publication Data
Crompton, T. R. (Thomas Roy)
The analysis of natural waters/T. R. Crompton.
p. cm.
Includes bibliographical references and indexes.
Contents: v. 1. Complex-formation preconcentration techniques—
v. 2. Direct preconcentration techniques.
1. Water—Analysis. 2. Water chemistry. I. Title.
TD380.C75 1993 628.1'61—dc20 92-23860

ISBN 0-19-855395-1 Vol. 1
ISBN 0-19-855394-3 Vol. 2
ISBN 0-19-855752-3 (2 vol. set)

Typeset by Colset Pte. Ltd., Singapore

Printed in Great Britain by
Bookcraft Ltd.,
Midsomer Norton, Avon.

PREFACE

Despite the great strides forward in analytical instrumentation that have been made in the last decade, the analyst working in the fields of potable water analysis and environmental analysis of non-saline and saline waters finds that, frequently, the equipment has insufficient sensitivity to be able to detect the low concentrations of organic and inorganic substances present in his samples with the consequence that he has to report less than the detection limit of the method. Consequently, trends upwards or downwards in the levels of background concentrations of these substances in the environment cannot be followed. This is a very unsatisfactory situation which is being made worse by the extremely low detection limits being set in new directives on levels of pollution, issued by the European Community and other international bodies. To overcome the problem, there has been a move in recent years, to apply preconcentration to the sample prior to analysis so that, effectively, the detection limit of the method is considerably reduced to the point that actual results can be reported and trends followed.

The principle of preconcentration is quite simple. Suppose that we need to determine 5 ng $litre^{-1}$ of a substance in a sample and that the best technique has a detection limit of 1 μg $litre^{-1}$ (1000 ng $litre^{-1}$). To reduce the detection limit to 5 ng $litre^{-1}$, we might, for example, pass 1 litre of the sample down a small column of a substance that absorbs the substance with 100 per cent efficiency. We would then pass down the column 5 ml of a solvent or reagent which completely dissolves the substance from the column thereby achieving a preconcentration of $1000/5 = 200$. Thus, if the detection limit of the analytical method without preconcentration were 1000 ng $litre^{-1}$, then with preconcentration it would be reduced to approximately 5 ng $litre^{-1}$.

The use of a column is but one of many possible methods of achieving preconcentration. Each chapter of the book discusses a different method of preconcentration and its application to the preconcentration of cations, anions, organic substances, and organometallic compounds. The book is based on a survey of the recent world literature dealing with preconcentration and its application to saline and non-saline waters.

A combination of preconcentration with the newest, most sensitive, and, by definition, most expensive analytical techniques now becoming available is achieving previously undreamt of detection limits at the very time when the requirements for such sensitive analysis is increasing at a rapid pace. Thus the combination of preconcentration with graphite furnace, Zeeman or inductively coupled plasma atomic absorption spectrometry and, particularly, the combination of the latter technique with mass spectrometry is enabling exceedingly low concentrations of metals in the ng $litre^{-1}$ or lower range to be determined. Preconcentration prior to gas or high performance liquid chromatography is achieving similar results in the analysis of organics.

Another aspect of preconcentration is, however, worthy of mention, particularly in the case of smaller laboratories which cannot afford to purchase the full range of modern analytical instrumentation. Using older, less sensitive instrumentation preconcentration will still achieve very useful reductions in detection limits which will be adequate in many but not all instances. Thus, if conventional atomic absorption spectrometry achieves detection limits of 1 and 5 mg $litre^{-1}$ for cadmium and lead in water, then a 200-fold preconcentration will reduce these limits to approximately 5 and 25 μg $litre^{-1}$ and a 1000-fold preconcentration will achieve 1 and 5 μg $litre^{-1}$.

This first volume of *The analysis of natural waters* covers those techniques which involve the formation of a complex of the species to be preconcentrated followed by concentration of the complex by techniques based on solvent extraction or adsorption on a column. This is followed by desorption with a small volume of a suitable reagent and then, in each case, by analysis using an appropriate analytical technique.

Throughout the book emphasis is laid on providing practical experimental detail so that the reader can, in many cases, apply the methods without reference to source literature and will be in a position to adopt procedures to his or her particular requirements. This is a field where much remains to be discovered and it is hoped that this book will assist chemists to further develop procedures.

Whilst the book has been written with the interest of water chemists in mind, many of the preconcentration procedures discussed could with little or no modification be applied to improving detection limits in laboratories in a wide range of other industries including, electronics, semiconductors, metallurgy, pharmaceuticals, organic chemicals, petroleum, and polymers.

The following groups of people will find much to interest them: management and scientists in all aspects of the water and other industries, river management, fishery industries, sewage effluent treatment and disposal, and land drainage and water supply, as well as management and scientists in all branches of industry which produce aqueous effluents. It will also be of interest to agricultural chemists; agriculturalists concerned with the ways in which chemicals used in crop and soil treatment permeate the ecosystem; to the biologists and scientists involved in fish, insect, and plant life; and to the medical profession, toxicologists, public health workers, and public analysts. Other groups of workers to whom the book will be of interest include oceanographers, fisheries experts, environmentalists, and, not least, members of the public who are concerned with the protection of the environment. The book will also be of interest to practising analysts, and, not least, to the scientists and environmentalists of the future who are currently passing through the university system and on whom, more than ever previously, will rest the responsibility of ensuring that by the turn of the century we are left with a worthwhile environmental to protect.

Anglesey T. R. C
July 1992

CONTENTS

CONTENTS OF VOLUME 2

1

INTRODUCTION

Many factors can affect the efficiency of a preconcentration procedure, and the more important of these are discussed below by means of illustrative examples concerning different preconcentration techniques. One very important factor to be borne mind when applying preconcentration procedures is the contribution made to the final analytical result by concentrations of the element to be determined present in the reagents used in the analysis. Obviously, this has an important bearing on the detection limit that can be achieved by preconcentration.

1.1 Purity of reagents and detection limits

Reagent purity is a very important consideration in all the preconcentration procedures discussed in both volumes of this book. It is considered below in the context of the chelation–solvent extraction procedure discussed in Chapter 2 (volume 1) but similar considerations apply to all preconcentration methods in which reagents are used.

Consider the preconcentration and determination of traces of a metal in water by the formation of an organic solvent soluble metal chelate prior to determination by, for example, atomic absorption spectrometry. In such procedures various reagents such as mineral acids, buffers, and organic complexing agents are added to a large volume of the aqueous sample in order to produce the metal chelates at a suitable pH. Once the chelate is formed it is extracted into a much smaller volume of an organic solvent such as freon, thereby achieving the necessary preconcentration. At this stage either the organic extract is analysed directly or an aqueous back extract into a small volume aqueous mineral acid is analysed. As well as the final extract containing a contribution of the metal to be determined originating in the sample there is also a contribution from the reagents used. As a rule of thumb, in these circumstances when the weight of metal contributed by the reagents equals the weight contributed by the aqueous sample then the detection limit of the method has been reached. When low detection limits are required then contributions of impurities from reagents must be kept as low as possible by using very pure reagents and, in some cases, by prepurifying some or all of the reagents immediately prior to using them in the analysis. Some idea of the variation in levels of impurities in commercial supplies of reagents can be obtained by considering the example of concentrated nitric acid.

Table 1 Specification analysis of concentrated nitric acid grades available from Rhone–Poulenc.

	Normatom ultra pure (65%)	Norma pure (69%) Analysis of Cd, Hg, Pb	Norma pure (69%) ISO–ACS reagent mg kg^{-1}	Norma pure (68%) RP	Rectapure (68%)	Reagent nitric acid (68%)
Ag	0.001					
Al	0.005					
As	0.001	0.01	0.01	0.01		
Au	0.010					
B	0.020					
Ba	0.005					
Be	0.001					
Bi	0.001					
Ca	0.010		1			
Cd	0.001	0.005	0.01			
Co	0.001					
Cr	0.001					
Cu	0.001		0.05			
Fe	0.010	0.2	0.2	1	10	
Ga	0.005					
Hg	0.001	0.001				

In	0.002					
K	0.020					
Li	0.005					
Mg	0.010					
Mn	0.001		0.4	0.4		
Mo	0.001					
Na	0.005					
Ni	0.001		0.05			
Pb	0.005	0.005	0.05	0.05		
Sn	0.005					
Sr	0.005					
Ti	0.001					
Tl	0.005					
V	0.001					
Zn	0.005		0.1			
Ce	0.050	0.005	0.5	0.5	10	
PO_4	0.010		2	2		
SO_4	0.[illegible]	1	1	2		
Price[a]	£12.20[b]	£24.60	£7.30	£6.20	£6.20	£3.70

[a] Price per litre (glass bottle) in 1989
[b] £29.56 per 250 ml (plastic bottle) at 00.00.92.

Table 2 Detection limits expected using various grades of concentrated nitric acid and preconcentrating factors.

Initial volume of aqueous sample (Vme)	Volume of sample extract	Preconcentration factor	Grade of nitric acid					
			Normaton ultrapure (65%) Element	LD[a] (ng litre^{-1})	Normapure (69%) (for analysis of Cd, Hg, Pb) Element	LD[a] (ng litre^{-1})	Normapure (69%) ISO–ACS reagent Element	LD[a] (ng litre^{-1})
5	5	1	Cd	282	Cd	1 410	Cd	2 820
			Hg	282	Hg	282	Hg	–
			Pb	1410	Pb	1 410	Pb	14 100
			As	282	As	2 820	As	2 820
			Cu	282	Cu	–	Cu	14 100
			Fe	2820	Fe	56 400	Fe	56 400
			Mn	282	Mn	–	Mn	112 800
			Ni	282	Ni	–	Ni	14 100
200	5	40	Cd	7.05	Cd	35.2	Cd	70.5
			Hg	7.05	Hg	7.05	Hg	–
			Pb	35.2	Pb	35.2	Pb	352.5
			As	7.05	As	70.5	As	70.5
			Cu	7.05	Cu	–	Cu	352.5
			Fe	70.5	Fe	1 410	Fe	1 410
			Mn	7.05	Mn	–	Mn	2 820
			Ni	7.05	Ni	–	Ni	352.5

1000	5	200	Cd	1.41	Cd	7.05	Cd	14.10
			Hg	1.41	Hg	1.41	Hg	–
			Pb	7.05	Pb	7.05	Pb	70.5
			As	1.41	As	14.10	As	14.10
			Cu	1.41	Cu	–	Cu	70.50
			Fe	14.1	Fe	282	Fe	282
			Mn	1.41	Mn	–	Mn	564
			Ni	1.41	Ni	–	Ni	70.5
5000	5	1000	Cd	0.28	Cd	1.41	Cd	2.82
			Hg	0.28	Hg	0.28	Hg	–
			Pb	1.41	Pb	1.41	Pb	14.10
			As	0.28	As	2.82	As	2.82
			Cu	0.28	Cu	–	Cu	14.10
			Fe	2.82	Fe	56.4	Fe	56.4
			Mn	0.28	Mn	–	Mn	112.8
			Ni	0.28	Ni	–	Ni	14.10

[a] From equation

$$\text{LD (ng litre}^{-1}) = \frac{2 \times M \times N \times 1.41 \times 10^6}{V} = \frac{2 \times M \times 0.5 \times 1.41 \times 10^6}{V} = \frac{1\,410\,000M}{V}$$

where M = concentration (mg kg^{-1}) of impurity in concentrated nitric acid (see Table 1); N = volume (ml) of concentrated nitric used in determination (0.5 ml); V = volume (ml) of aqueous sample taken for preconcentration.

In Table 1 are given the specification analyses of a range of nitric acids obtainable from Rhone–Poulenc. These range from the Normatom ultra pure grade where analyses are quoted for 26 elements in the 0.001–0.005 mg kg^{-1} (1410–7050 ng $litre^{-1}$) range and eight elements in the 0.01–0.1 mg kg^{-1} (14 100–141 000 ng $litre^{-1}$) range.The intermediate Normapure ranges of concentrated nitric acid specify the analyses in the case of the ISO–ACS reagent quality 12 elements only occuring in the range 0.01–2 mg kg^{-1} (14 100–2 820 000 ng $litre^{-1}$). The Rectapure ordinary grade specifies the analysis of only two impurities, both present at the 10 mg kg^{-1} level. Clearly, then the Normatom grade of acid is the only one worth considering for use in preconcentration techniques. Similar considerations will apply to any other reagents used in the analysis.

Consider, as an example, the preconcentration and determination of a metal by a chelation–solvent extraction procedure of the type discussed above. To the sample volume (up to 1 litre) is added N ml of concentrated nitric acid of specific gravity 1.41 containing M mg kg^{-1} of copper as impurity. The weight (W) of the metal in V ml of concentrated acid is therefore:

$$W(\text{ng}) = M \times N \times 1.41 \times 10^3$$

In order for detection of the element to be achieved the weight of the element in V ml of sample should be $2W$ (ng) $= 2 \times M \times N \times 1.41 \times 10^3$ ng per V ml of sample, i.e. $(1000 \times 2 \times M \times N \times 1.41 \times 1000)/V$ ng element per litre which is taken as the detection limit (LD) for the element concerned, i.e.

$$\text{LD} = \frac{2\,820\,000MN}{V}$$

where M is impurity concentration (mg kg^{-1}) in concentrated nitric acid; N = volume (ml) of concentrated nitric acid used in determination; V = volume (ml) of aqueous sample taken for preconcentration.

If 0.5 ml of concentrated nitric acid us used then

$$\text{LD} = \frac{1\,410\,000M}{V}$$

If the concentration of the element in the concentrate of nitric acid is 0.001 mg kg^{-1} (M) and 1 litre sample (V) is taken for analysis

$$\text{LD} = \frac{1\,410\,000 \times 0.001}{1000} = 1.41 \text{ ng litre}^{-1}$$

In Table 2 are tabulated the expected limits using three grades of nitric acid of different purities when 5 ml, 200 ml, 1000 ml, and 5000 ml

Table 3 Grades of acetone available from Rhone–Ponlenc.

Grade	Non-volatile residues at 100 °C (mg kg^{-1})	Acidity as CH_3COOH (mg kg^{-1})
Acetone for pesticide analysis	3	–
Acetone for HPLC	5	18
Acetone for UV spectroscopy	5	–
R P Normapure A R	10	20
Acetone Rectapure	50	–
Acetone, laboratory grade	–	–

aqueous sample are preconcentrated to 5 ml of final extract, i.e. preconcentration factors, respectively of ×1 (no preconcentration), ×40, ×200, and ×1000. Taking cadmium as an example it is seen that detection limits obtainable with ultra pure Normatom nitric acid range from 282 ng $Litre^{-1}$ (no preconcentration) to 0.28 ng $Litre^{-1}$ (×1000 preconcentration). For the Normapure acids detection limits for cadmium are much poorer ranging from 1410 ng $Litre^{-1}$ (no preconcentration) to 1.41 ng $Litre^{-1}$ (×10000 preconcentration). Clearly then, if a detection of down to 2 ng $Litre^{-1}$ of cadmium is required then Normaton nitric acid must be used and a preconcentration factor of 200 adopted, e.g. 1 litre of sample preconcentrated to 5 ml of extract.

Similar considerations apply in the preconcentration and determination of organic substances. Thus Rhone–Ponlenc quote six grades of acetone ranging in non-volatile residue contents from 3 to 50 mg kg^{-1}, (Table 3). Clearly, if, for example, pesticides are being preconcentrated and acetone is the solvent then the pesticide analysis grade is preferable to the others.

1.2 Prepurification of reagents

If reagents of adequate purity to meet detection limit requirements in preconcentration methods are to be available then calaboratory purification of reagents may be necessary. Indeed many workers prefer to clean up their own reagents to stipulated degrees of purity so that they can achieve better control on the environmental contamination factors controlling the results obtained in analyses. In this approach some or all of the reagents are purified immediately prior to their use in the analyses or as part of the analytical procedure. A good example of this is the cleaning up of ammonium pyrrolidinedithiocarbamate–diethylammonium dithiocarbamate mixed chelating reagent before its use in the preconcentration of nickel, copper, cadmium, and lead.

In this procedure 1 g of each chelating agent is dissolved in 50 ml water in a separatory funnel. Freon (1,1,2-trichlorotrifluoroethane) (100 ml) is added. After agitation the lower freon layer containing metallic impurities is separated and rejected, and a similar extraction with 100 ml Freon performed. The resulting aqueous solution of chelating agents is now virtually free from metallic contamination and is suitable for use in the preconcentration of samples.

1.3 Effect of pH

In a typical chelation–solvent extraction procedure (discussed in Chapter 2, volume 1), a small volume of an aqueous solution of ammonium pyrolidinediethyldithiocarbamate is added to a large volume of aqueous sample, and the metal chelates extracted with a relatively small volume of methyl ethyl ketone.[1,2] The pH of the aqueous sample has a pro-

Table 4 Influence of pH on extraction of metal pyrrolidine diethyldithriocarbamates from water with methyl ethyl ketone.

Element	Element recovery
Co, Cu, Fe, Ni, Mn, Zn	100% at pH 4.0–4.6
Cd, Pb	100% at pH 4.0–4.6 but complex only stable 2–3
Ag	100% at pH 1–2
Cr	100% at pH 1.8–3.0
Mo	100% at pH 1–1.5

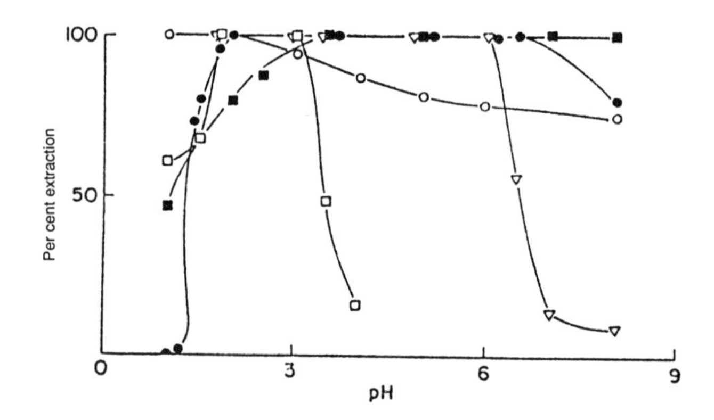

Fig. 1 Effect of pH on the extraction of some trace metals using the APDC-MIBK procedure (aqueous/organic = 5); ○ Ag, 4 μg litre^{-1}; □ Cr, 20 μg litre^{-1}; ▽ Fe, 20 μg litre^{-1}; ● Mn 6 μg litre^{-1}; ■ Pb, 8 μg litre^{-1}. From Reggeks and Van Grieken[3] with permission.

found effect on the efficiency with which various metals are extracted as illustrated in Table 4 and Figs 1 and 2. Large decreases in recovery were observed for all metals at pH values below 3.0 and small decreases at pH values greater than 8.0.

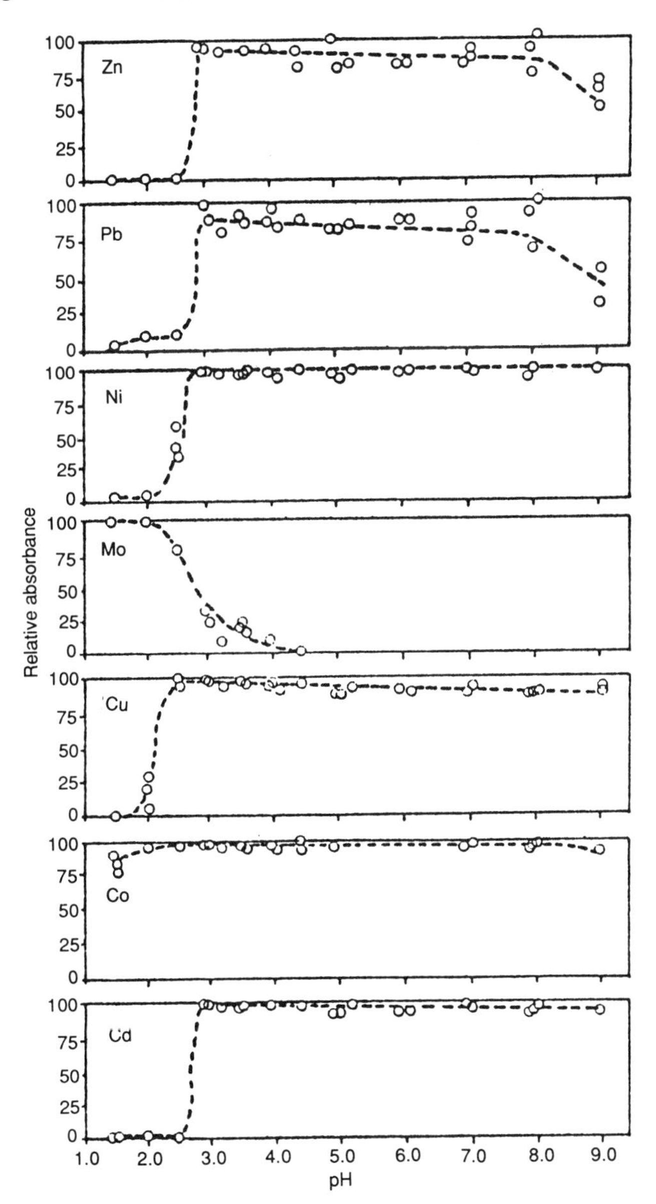

Fig. 2 Efficiency of the APDC/MIBK method as a function of sample pH. From Subramanian and Meranger[1] with permission.

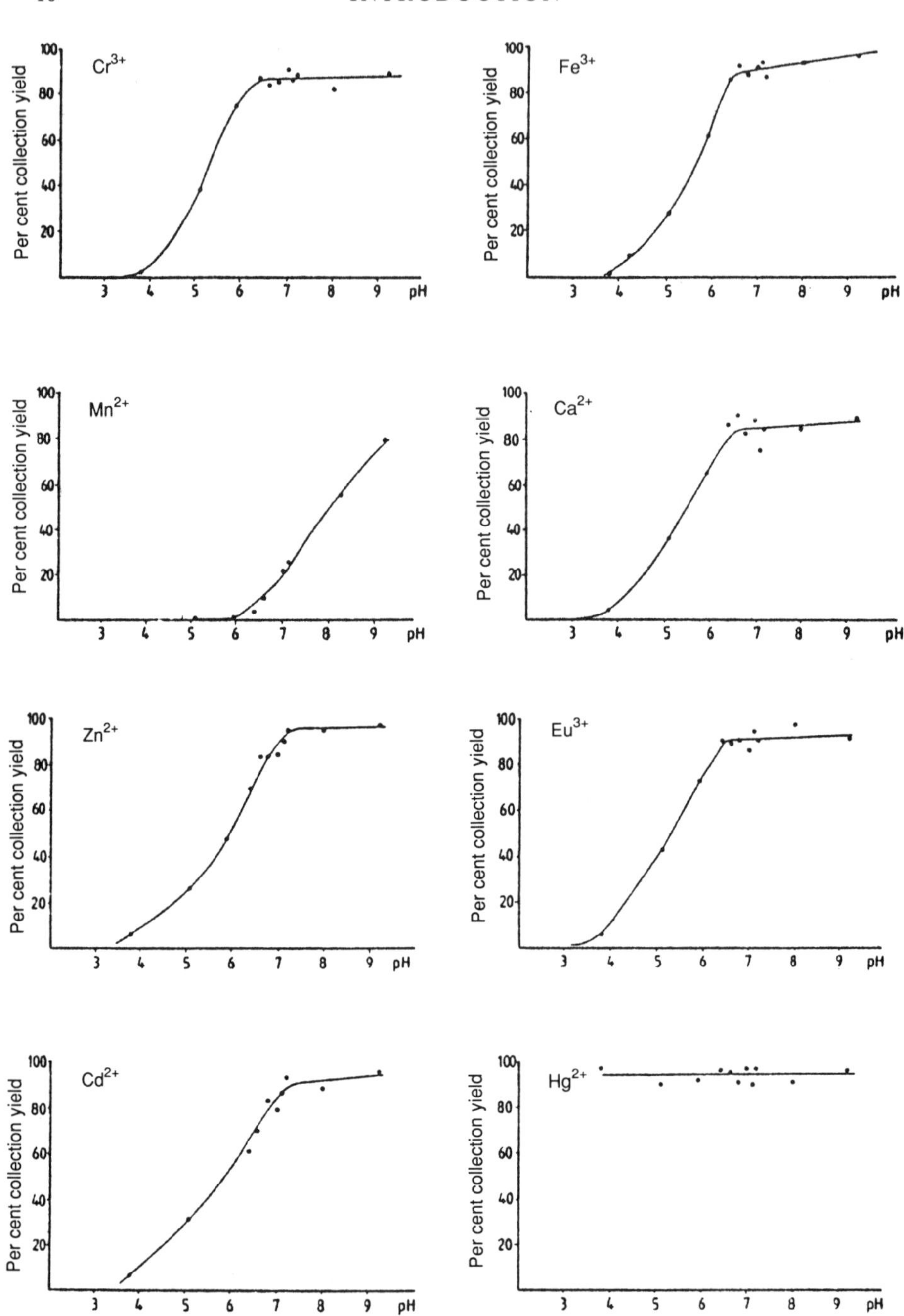

Fig. 3 Influence of the pH on the collection efficiency of cellulose DEN powder for Cr^{3+}, Mn^{2+}, Fe^{3+}, Ca^{2+}, Zn^{2+}, Cd^{2+}, Eu^{3+}, and Hg^{2+}. From Reggeks and Van Grieken[3] with permission.

2,2-Diaminodiethylamine cellulose powder has particularly good chelation properties for metals achieving a chelation capacity of 1.5 mol equivalent g^{-1} resin.[3] Figure 3 illustrates the effect of pH on the metal collection efficiency of this resin. At pH 7, 90–100 per cent recovery is obtained for chromium, zinc, cadmium, iron, cobalt, europium, and mercury whilst only a 30 per cent recovery of manganese is obtained. In all cases recoveries decrease sharply at pH values below 6.

1.4 Effect of chelating agent–metal ratio

Another factor which affects recovery in chelation–solvent extraction procedures of the type discussed in Chapter 2, Volume 1, is the ratio of the molar concentration of the chelating agent to the molar concentration of metal present in the aqueous sample. The results in Fig. 4 illustrate that in the methylethyl ketone extraction of metal pyrrolidinedithiocarbamates from water samples 100 per cent recovery of cadmium, lead, silver, nickel, and copper is obtained when the chelating agent:metal ratio (log (APDC/metal)) is 3.0 whilst a ratio of 5.0 is required to obtain full recovery of iron, chromium, and cobalt.[4] It is necessary, therefore, when developing such procedures to carry out a preliminary systematic examination of the effect of molar excess of chelating agent used on metal recovery in the preconcentration step.

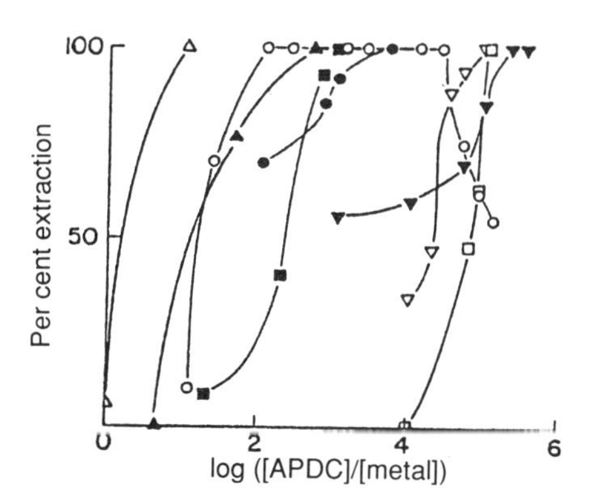

Fig. 4 Effect of APDC on the extraction of some trace metals using the APDC-MIBK procedure: ▲ Ag, 4 μg $litre^{-1}$; ● Cd, 0.2 μg $litre^{-1}$; ▽ Co, 20 μg $litre^{-1}$; ▼ Cr, 20 μg $litre^{-1}$; △ Cu, 20 μg $litre^{-1}$; □ Fe, 20 ng ml^{-1}; ■ Ni, 50 μg $litre^{-1}$; ○ Pb, 8 μg $litre^{-1}$. From Burba and Willman[4] with permission.

1.5 Matrix interference effects

Water samples might contain either naturally occurring chelating agents (e.g. humic acids) or man made chelating agents (e.g. nitriloacetic acid or linear alkylbenzene sulphonates) that form strong complexes with metals present in the samples. Such complexes might be stronger than the complex formed between metals and the chelating agent added to the sample in chelation–solvent extraction methods of preconcentration which would under these conditions give low metal recoveries. Such effects can be overcome by using a standard addition method for calculation or by giving the water sample a preliminary treatment with agents such as ozone or by exposure to ultraviolet light to decompose such complexes before proceeding with the preconcentration.

Another type of matrix effect is covered by the presence in the sample of relatively high concentrations of inorganic solids such as occurs, for example, in the case of sea or estuary water samples. Such an effect is illustrated in Fig. 5 which shows the effect of the concentration of sodium chloride on the partition coefficient of metals between the water sample, adjusted to pH 11, and cellulose. In the absence of sodium chloride K_d is in the range 500–2000 for iron, nickel, cadmium, copper,

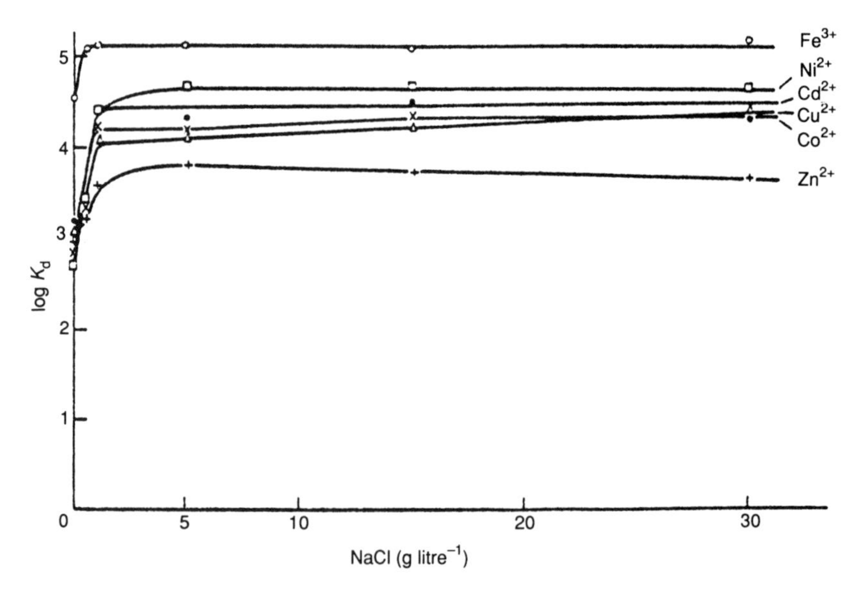

Fig. 5 Influence of electrolyte solution concentration (0–30 g litre^{-1} NaCl) on heavy metal sorption (Cd^{2+}, Co^{2+}, Cu^{2+}, Fe^{3+}, Ni^{2+}, Zn^{2+} each 20 μg litre^{-1}) on cellulose. From Burba and Williams, [4] with permission.

cobalt, and zinc. The presence of 1.5–2.0 g $litre^{-1}$ of sodium chloride enhances K_d by a factor of 30 to 15 000–60 000. This indicates that the deliberate addition of salt to an aqueous sample might have beneficial effects in preconcentrations which involve adsorption of metals on to an adsorbent, followed by desorption with a small volume of a reagent.

1.6 Adsorption efficiency of solid adsorbents

The efficiency of adsorption of organic substances from water on to columns of microreticular resins (Chapter 3, Volume 1) is influenced by several factors such as the flow rate of the water through the column and sample pH. The subsequent desorption of the preconcentrated organic, by either extraction with a relatively small volume of organic solvent such as acetone or carbon disulphide or thermally, can be affected by desorption time and temperature. In a typical thermal preconcentration technique[5] based on these principles the water sample is passed down a column of XAD-2 or XAD-4 resin at a controlled flow rate. At the end of this stage the XAD-2 column is connected to a small column of Texax GC and the organics swept on to the latter column with a purge of helium at 230 °C. The Texax GC column is then heated to 45 °C and swept with helium to remove residual water. Finally, the Texax GC column is heated to 220 °C and purged with nitrogen on to a gas chromatograph or gas chromatograph–mass spectrometer for quantification and identification.

In a typical solvent extraction preconcentration technique the water is passed down a column of XAD-2 or XAD-4 resin and then a relatively small volume of carbon disulphide or acetone is passed down the column to elute organics prior to gas chromatographic analysis. Some typical recovery data for various types of organic compounds from macroreticular resins are given in Table 5. As might be expected adsorption efficiency of the organic from the water sample on to the resin and the desorption efficiency for removal of the organic from the resin either by solvent extraction or thermally differ appreciably from one type of organic compound to another. When devising such methods for the preconcentration of particular organic compounds carefully controlled and systematic recovery studies should be conducted to check on recovery and on reproducibility of recovery in order to establish experimental conditions for maximum recovery. If recoveries are acceptably reproducible but low, then the preconcentration step should be included in the method calibration procedure adopted.

A further method of metal preconcentration (discussed in Chapter 2, Volume 2) involves passage of the aqueous sample through a bed of a metal oxide, usually manganese dioxide, titanium dioxide, zirconium

Table 5 Adsorption efficiency of organics on macroreticular resins.

Organic	Per cent recovery carbon disulphide desorption		Per cent recovery thermal desorption	
	200 ml water sample containing 2–10 $\mu g\ litre^{-1}$ of organic	200 ml water sample containing 100 $\mu g\ litre^{-1}$ of organic	20 ml of water sample containing 10 $\mu g\ litre^{-1}$ of organic	200 ml of water sample containing 200 $\mu g\ litre^{-1}$ of organic
Aromatic hydrocarbons	72 ± 7	76 ± 11	87 ± 8	90 ± 11
Polyaromatic hydrocarbons	—	—	88 ± 3	85 ± 2
Aliphatic chloro hydrocarbons	90 ± 2	56 ± 1	—	—
Aromatic halo hydrocarbons	94 ± 12	—	88 ± 3	84 ± 7
Alcohols	91 ± 8	72 ± 18	98 ± 6	89 ± 7
Ketones/aldehydes	99 ± 3	98 ± 6	97 ± 2	84 ± 7
Esters	78 ± 8	50 ± 18	97 ± 2	87 ± 10
Phenols	—	—	84 ± 20	77 ± 10

From Ryan and Fritz[6] with permission.

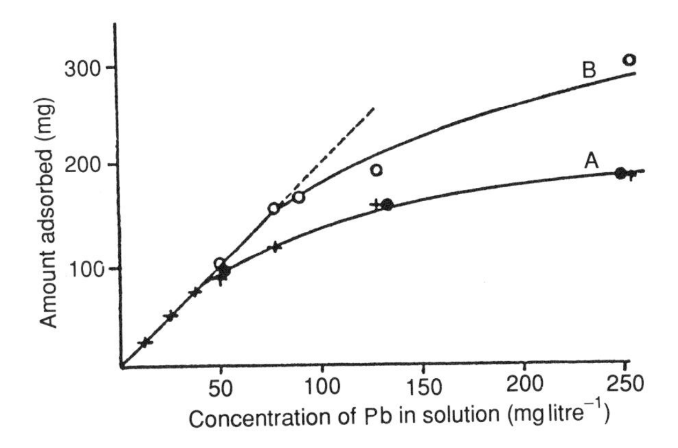

Fig. 6 The amount of lead adsorbed from solution versus amount in solution. Curve A; Potable water (+) and sea water (•), 11 cm diameter filter. Curve B, potable water, 4.7 cm filter system. The dashed line represents quantitative (100 per cent) adsorption from solution. From Matthews[7] with permission.

dioxide, alumina, hydrated iron oxide or silica or C_{18}-bonded silica. The adsorbed metal is then desorbed with a volume of aqueous sample taken in order to achieve preconcentration. Generally speaking, such preconcentrations are carried out on relatively thin beds rather than columns of the metal oxide. In Fig. 6 (curve A) is shown the relationship between the weight of lead adsorbed on a 0.25 cm thick disc of manganese dioxide from 2 litres of saline and non-saline aqueous solutions containing between 10 and 500 mg lead, i.e. (5–250 mg Litre^{-1} lead).[7] The flow rate used in this experiment was 500 ml min^{-1}, i.e. 4 min to pass through a 2-litre sample. It is seen that 100 per cent adsorption of lead on manganese dioxide occurs when the sample contains up to 38 mg litre^{-1} lead (75 mg lead), and then drops dramatically to 50 per cent recovery when the sample contains 150 mg litre^{-1} lead (300 mg lead). Increasing the thickness of the manganese dioxide layer eightfold from a 0.25 cm to a 2 cm thick disc and using the same sample flow rate extended the range over which quantitative recovery of lead fron non-saline and saline water samples was obtained, from 38 mg litre^{-1} (75 mg lead) to 80 mg Litre^{-1} (160 mg lead). Increasing the pH from 2.0 to 7.4 did not affect recovery.

In such concentration methods the factors which most affect recovery are the flow rate of the aqueous sample and the thickness and possibly area of the collecting bed. Such parameters should be carefully studied when devising procedures.

1.7 Ion exchange resin theory

This is discussed further in Chapters 4 and 5 (Volume 1). When an ion exchange reaction is carried out by a 'batch' method – that is, by putting a quantity of resin into a certain volume of solution – the reaction begins at once, but a certain time elapses before the equilibrium state is reached. It is generally a simple matter to determine the rate of exchange, for example, by sampling and analysing the solution at intervals or by making use of some physical property, such as electrical conductivity, which changes as the reaction proceeds. Such experiments show that rates of exchange vary very much from one system to another, the times of half-exchange ranging from fractions of a second to days or even months in certain extreme cases. Figure 7 shows the progress of exchange with a number of different cations exchanging with equal samples of the ammonium form of a phenolsulphonic acid resin under the same conditions.

Clearly, a knowledge of the rate of exchange is a prerequisite for the most effective use of resins. The factors which influence the rate of exchange and show how they account quantitatively for the form of kinetic curves such as those shown in Fig. 7 are listed below. By a series of controlled experiments in which one factor is varied at a time, it can readily be shown that, other things being equal, a high rate of exchange is generally favoured by the following choice of conditions:

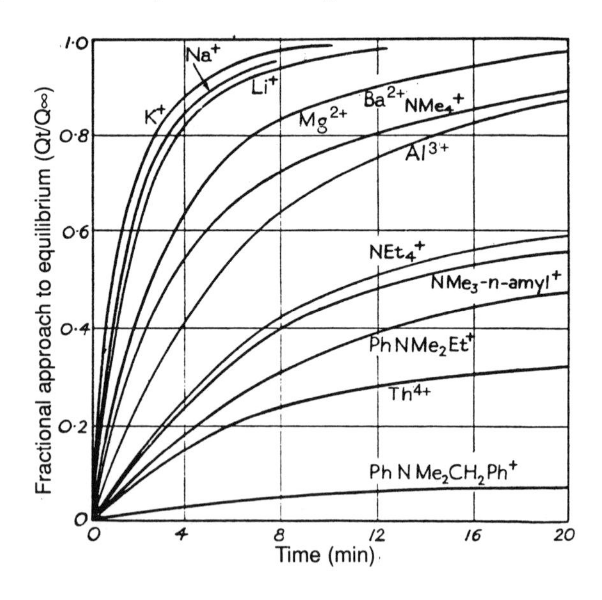

Fig. 7 Rate of exchange of phenolsulphonic acid resin (in ammonium form) with various cations.

(1) a resin of small particle size;
(2) efficient mixing of resin with the solution;
(3) high concentration of solution;
(4) a high temperature;
(5) ions of small size;
(6) a resin of low cross-linking.

The exchange reactions are generally much slower than reactions between electrolytes in solution, but there is no evidence to indicate the intervention of a slow 'chemical' mechanism such as is involved in most organic reactions, where covalent bonds have to be broken. The slowness of exchange reactions can be satisfactorily accounted for by the time required to transfer ions from the interior of the resin grains to the external solution, and vice versa. The rate of exchange is therefore seen to be determined by the rate at which the entering and leaving ions can change places. The process is said to be 'transport-controlled'; it is analogous, for instance, to the rate of solution of a salt in water. All the factors (1) to (6) listed above are such as to facilitate the transport of ions to or from or through the resin.

In preconcentration techniques the batch procedure, i.e. putting a quantity of resin into a certain volume of solution, is not usually used. Most practical applications of ion exchange resins in preconcentration involve column processes wherein a large volume of sample is passed down a small column of resin. For example if cation B is to be preconcentrated from water by means of a sodium form resin,

$$\text{NAR} + \text{B}^+(\text{aq}) \rightarrow \text{BR} + \text{Na}^+(\text{aq})$$

it would be inefficient to shake a quantity of resin with a given volume of water, as an equilibrium would be set up, and some of the cation B would be left in solution. Instead the water is passed through a column of the resin, and as the water becomes depleted of cation B by contact with the first layers, it passes to fresh resin containing little or no cation B, and so the equilibrium is constantly displaced. With a long enough column the effluent is entirely free from cation B.

In order to arrange conditions for efficient operation of columns some regard must be paid to the 'local' kinetics of the exchange reactions. The quantitative theory[8] of column processes is somewhat specialized, but a qualitative understanding of the principal factors is readily obtained.

The usual methods of operating columns can be divided into (A) displacement, in which the ion on the column is sharply displaced by another, more strongly sorbed, and (B) elution, in which the ion is more gradually moved down the column by treatment with a more weakly sorbed element. (A) leads to sharp bands travelling at a rate determined

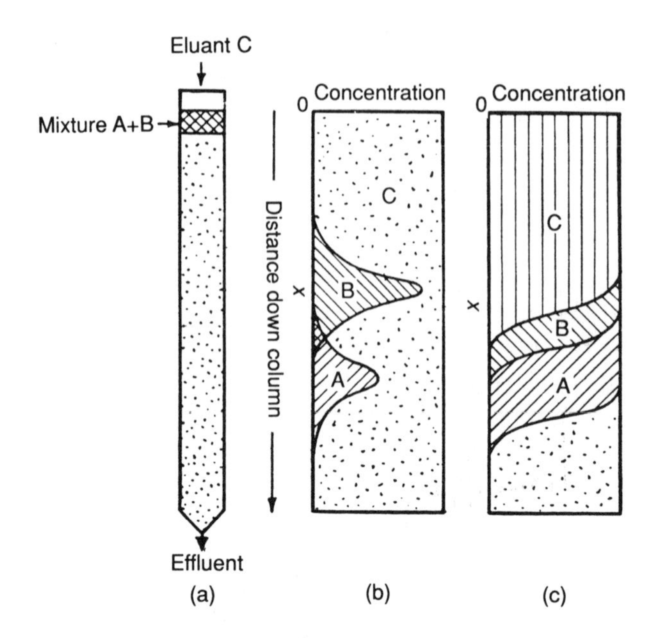

Fig. 8 Form of bands in elution (b) and displacement (c) chromatography: (a) represents the column, x distance down column, and figures (b) and (c) show the concentrations of ions A, B, and C along the column after a certain time.

by the flow of incoming solution. (B) leads to bands with somewhat diffuse fore and aft boundaries, travelling at a rate dependent on the relative affinity of the resin for the two ions. (A) is useful for dealing with large quantities ('heavy loading' of the column), while (B) is the preferred process for obtaining optimum separation of substances, albeit in small quantities. Figure 8 shows diagrammatically the types of band obtained with a ternary mixture of A, B, and C in the two cases.

Whatever the purpose of the process to be carried out on the column—absorption, displacement or chromatographic analysis or separation—the highest efficiency would be reached if the liquid passing down the column came to full equilibrium with each layer of resin grains. This condition, however, would require an infinitesimal rate of flow. At all practical rates of flow non-equilibrium conditions prevail with reduced efficiency. The essence of column operation is to choose an appropriate compromise between efficiency and speed, compatible with the required result.

The factors militating against attainment of local equilibrium are:

(a) non-uniformity of the solution in the interstices, i.e. film-diffusion, resistance;

(b) non-uniformity within the grains of resin, i.e. particle-diffusion resistance; and

(c) irregular flow of the solution down the column, spoiling sharpness of the moving bands of solute.

The obvious steps to be taken in the direction of improving efficiency are:

(1) the use of resins of small particle size;

(2) low flow-rate;

(3) an elevated temperature (to increase diffusion coefficients); and

(4) great care in the packing of the column;

but these are offset by the extra pressure, time or experimental difficulty incurred.

Assuming that factor (c) could be ignored (which, however, is practically never the case) it is possible to calculate the performance of a column under conditions where film- or particle-diffusion are rate determining.[8] A more practicable approach, however, which covers any form of non-equilibrium, is to use a semi-empirical treatment similar to the 'plate theory' employed in the theory of distillation columns.[9,10] According to the plate theory, the column can be considered as consisting of a number (N) of sections ('theoretical plates') in each of which the average concentration (c) of solution in the pores can be considered as effectively in equilibrium with the average amount of solute sorbed by the resin. The effectiveness of the column can be judged by the number of theoretical plates it appears to contain, and once N is known the elution curve can be predicted.

Consider by way of example, the most important case of the application of the plate theory—the separation of two similar species, A and B, by elution chromatography with an eluant, C. On preconcentration it may be required, for example, to retain species A on the column (i.e. the preconcentrated species) and to allow mobile species B to be completely eluted from the column (as this substance might interfere in the determination of A in the final analytical step on the eluted preconcentrate). A small quantity of the mixture of A + B is first sorbed on the top of the column (previously in form C), and then 'developed' by elution with the solution of C. As the mixed band moves down the column, A and B gradually separate or in the ideal case species A hardly moves and species B moves relatively quickly down the column, because of their different sorption affinites. If the 'loading' of the column is light, the sorption of A and B can be considered linear with concentration, and each is characterized by an equilibrium distribution coefficients, K_d, defined by

$$K_d = \frac{\text{conc. of the ion in resin (mol equiv. per g)}}{\text{conc. of the ion in solution (mol equiv. per ml)}}$$

The rate of movement of the band down the column is inversely proportional to K_d, and, hence, the rate of separation of A and B depends on the ratio of their distribution coefficients, $(K_d)_A/(K_d)_B$, which can be determined by simple batch equilibrium experiments, using the appropriate solution of C as the medium in excess. For a high rate of movement K_d is low and for a low rate of movement K_d is high, so to separate substance B with a high rate of movement from substance A with little or no movement, $(K_d)_A$ should be high and $(K_d)_B$ low, i.e. $(K_d)_A/(K_d)_B$ should be high.

The effective number of theoretical plates, N, in the column (which depends on the flow-rate) can be determined by a study of the elution curve for a suitable species (say A). An elution curve is a graph of the concentration (c) of the species in the effluent flowing from the column, as a function of the volume (v) of elutriant passed through. According to Glueckauf's theory[8,10] the shape of the elution curve is given approximately by

$$c = c_{max} \exp\left[-\frac{N}{2}\frac{(v_{max} - v)^2}{v v_{max}}\right] \tag{1.1}$$

(where c_{max} and v_{max} are the co-ordinates of the centre of the band); the position of the maximum is given by

$$c_{max} = \frac{m}{v_{max}}\sqrt{\frac{N}{2\pi}} \tag{1.2}$$

where m is the total quantity (mol equiv.) of the given solute in the band and is equal to the area under the elution curve; and the band width (at the concentration level $c_{max}/e = 0.368c_{max}$) is given by

$$\text{band width} = \frac{64v_{max}}{N^2} \tag{1.3}$$

N can be determined conveniently from either eqn (1.1) or (1.3). If two similar solutes, A and B, are being separated, N will be common to both, but v_{max} will be approximately inversely proportional to K_d, i.e. $(v_{max}K_d)_A = (v_{max}K_d)_B$.

If the distribution coefficients are rather similar, there will be a significant overlap of the two bands, and the purity of the fractions can be calculated. For equal purity of the separated bands, the 'cut' should be made at the volume given by

$$v = \sqrt{{}_A v_{max}\,{}_B v_{max}} + \frac{2\,{}_A v_{max}\,{}_B v_{max}(m_A^2 - m_B^2)}{N({}_B v_{max} - {}_A v_{max})(m_A^2 + m_B^2)}$$

where m_A and m_B are the total quantities of A and B present. The proportion of impurity of A in B (or vice versa) is given by

$$\frac{\Delta m_A}{m_B} = \frac{2m_A m_B}{m_A^2 + m_B^2}\left[\frac{1}{2} - S\left(\frac{\sqrt{N}\left(\sqrt{_B v_{max}} - \sqrt{_A v_{max}}\right)}{\sqrt[4]{_A v_{max}\, _B v_{max}}}\right)\right]$$

where S is the area under the tail of the overlapping curve, and is given by the appropriate integral of the normal errors curve (obtained from tables), i.e.

$$S = \frac{1}{\sqrt{2\pi}}\int_0^x \exp - \left(\frac{x^2}{2}\right)^2 dx, \quad \text{with } x = \frac{v_{max} - v}{\sqrt{v_{max} v}}\sqrt{N}$$

Since N is proportional to the length of the column, the improvement of separation obtainable by increasing the length can be calculated. Increase of cross-section of the column increases its handling capacity.

Glueckauf[10] has provided a chart showing the purity obtainable with different separation factors, $(K_d)_A/(K_d)_B$, and different numbers of plates; for example with a separation factor of 1.2 and a column of 1 000 plates, a purity of 99.9 per cent is obtainable (starting with equal quantities of A and B).

This brief explanation of the plate theory should suffice to show how a few well-conceived exploratory measurements make it possible to plan a critical preconcentration step. These factors which have a profound effect on the binding capacity of ion exchange resins include the pH of the aqueous sample and the presence of natural chelating agents such as humic acid as discussed previously.

1.8 Coprecipitation techniques

An example of this technique (discussed in Chapter 4, Volume 2) is the preconcentration of metals by the addition of iron followed by pH adjustment to an alkaline pH by addition of sodium hydroxide. The mixture is left for an equilibrium period during which the iron precipitates as hydroxide with some or all of the oxides of the metals to be preconcentrated occluded on to it. Following filtration of the combined metal oxides, they are dissolved in a relatively small volume of an acidic reagent and this preconcentrated solution analysed. One of the most important factors affecting recovery to be controlled during such preconcentrations is the pH to which the solution is adjusted with sodium hydroxide.

In Figure 9 is shown metal recovery–pH curves obtained in the preconcentration of manganese, nickel, copper, zinc, and lead[11] in 200 ml aqueous sample to which had been added 2 mg of ferric iron. Lead and

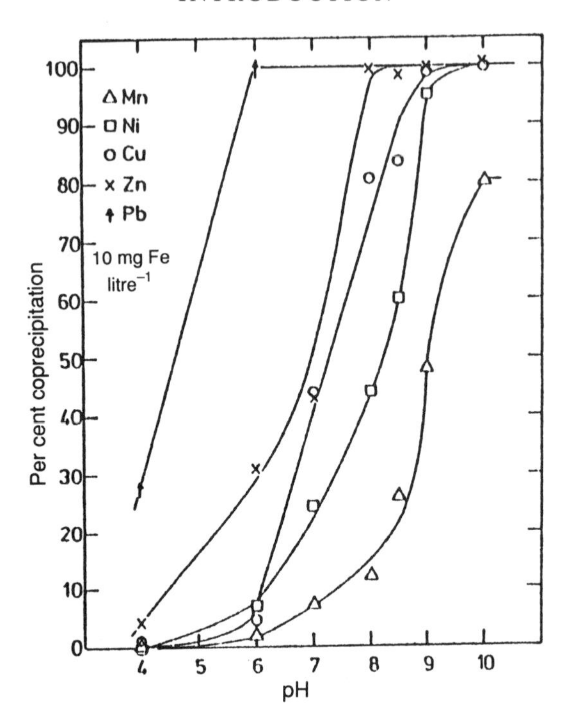

Fig. 9 pH dependence on the coprecipitation of Mn, Ni, Cu, Zn, and Pb in sea-water. From Chakrovarty and Van Grieken[11] with permission.

zinc are fully coprecipitated at a pH above 7–8, copper and nickel require a pH above 9, whilst manganese is only 60 per cent coprecipitated at a pH in excess of 10. Therefore, with the exception of manganese the optimum pH is 9. Using this procedure in conjunction with an analytical finish involving X-ray fluorescence spectrometry enabled preconcentration factors of up to 1.5×10^4 to be achieved.

1.9 Preconcentration of organics by direct solvent extraction

The following two factors affect the efficiency of such preconcentration procedures (discussed in Chapter 5, Volume 2).

(a) *Mutual solubility effects*. When a water sample containing a solute to be preconcentrated is shaken with a relatively small volume of organic solvent then at the end of the extraction the organic phase contains dissolved water and the water phase contains dissolved solvent, low recovery of solute in the organic extract will be obtained depending on the solubility of solute containing water in the solvent. Recoveries can be overcome by carrying out two or three successive solvent extractions with fresh portions of solvent, but at the expense of decreasing the preconcentration factor achieved, hence the detection limit of the overall procedure.

(b) *Partition effects*. When an aqueous phase containing a dissolved solute is shaken with an organic solvent, then, depending on the partition coefficient of the solute under the particular experimental conditions the solute will redistribute itself between the two phases. It is necessary to choose solvents which have a high preference for the solute.

The theory of these two effects is discussed below.

1.9.1 Theory of mutual solubility effects

If we start with V_w ml of water sample containing A μg of solute to be preconcentrated (i.e. $(A \times 1000)/V_w$ μg litre^{-1} solute in water) and V_s ml of organic solvent, then, assuming complete transfer of solute from water to solvent and taking into account the solubilities of solvent in water and water in solvent, the volumes of solvent and water at the end of the extraction are:

$$\left(V_w + \frac{V_w S_2}{100} - \frac{V_s S_1}{100}\right) = V_w\left(1 + \frac{S_2}{100}\right) - \frac{V_s S_1}{100} \text{ ml water}$$

$$\text{and } \left(V_s + \frac{V_s S_1}{100} - \frac{V_w S_2}{100}\right) = V_s\left(\frac{1 + S_1}{100}\right) - \frac{V_w S_2}{100} \text{ ml solvent}$$

where S_1 is the per cent v/v solubility of water in solvent and S_2 is the per cent v/v solubility of solvent in water.

The term $V_w S_2/100$ represents the volume of solvent still containing solute dissolved in the aqueous phase, as V_s ml of solvent contains A μg solute, $V_w S_2/100$ ml of organic solvent contains $A\ V_w S_2/100\ V_s$ μg solute. Therefore at the end of the first extraction we have $A\ V_w S_2/100\ V_s$ μg solute dissolved in $[V_w(1 + S_2/100) - V_w S_2/100]$ ml water, and $A - A\ V_w S_2/100\ V_s = A(1 - V_w S_2/100\ V_s \quad [V_s(1 + S_1/100) - V_w S_2/100]$ ml solvent.

It now remains to extract the $A\ V_w S_2/100\ V_s$ μg residual solute from the water phase by carrying out a second extraction with a further V_s ml of fresh organic solvent. The water obtained at the end of the first extraction is now already saturated with solvent and, hence, does not dissolve further solvent in the second extraction. The V_s ml organic solvent used in the second extraction is not saturated with water and dissolves $S_1\ V_1/100$ ml water. The volumes of solvent and water at the end of the second extraction are:

Table 6 Mutual solubility effects in extraction of solutes from water with organic solvents.

Volume (ml)		Original weight (μg) of solution aqueous phase	Solubility (per cent) of solution in aqueous phase	Weight of solute μg in organic phase	per cent recovery of solute	Preconcentration factor	
Water V_w	Organic solvent V_s	A	S_2			One extraction	Two extractions
A. 1 litre of water extracted with 10 ml solvent							
First extraction				$A\left(1-\frac{V_w S_2}{100V_s}\right)$	$A\left(1-\frac{V_w S_2}{100V_s}\right)\frac{100}{A}$	$\frac{V_w}{V_s\left(1+\frac{S_1}{100}\right)-V_w\frac{S_2}{100}}$ assuming $S_1 = 0.1$ per cent	$\frac{V_w}{2V_s\frac{(1+S_1)}{100}-\frac{V_w S_2}{100}}$
1000	10	20	0.05	19.0	95	105	
			0.1	18.0	90	111	—
			0.5	10.0	50	200	
Second extraction				$\frac{AV_w S_2}{100V_s}$	$\frac{100AV_w S_2}{100V_s A}$		
	10	—	0.05	1.0	5		
			0.1	2.0	10	—	—
			0.5	10.0	50		
Combined first and second extractions				$A\left(1-\frac{V_w S_2}{100V_s}\right)$	$+\ \frac{AV_w S_2}{100V_s} = A$		
			0.05	20	100		51
			0.1	20	100	—	52
			0.5	20	100		66

B. 1 litre of water extracted with 1 ml solvent							
First extraction							
1000	1	20	0.05	10	50	1996	–
			0.1	Nil[a]	–	–	
Second extraction							
	1	–	0.05	10	50	–	–
			0.1	Nil[a]	–		
Combined first and second extracts							
			0.05	20	100		666
			0.1	Nil[a]	–	–	
C. 10 litres of water extracted with 5 ml solvent							
First extraction							
10000	10	20	0.02	16.0	80	1250	–
			0.05	10.0	50	2000	
			0.01	Nil[a]	Nil	–	
Second extraction							
	10		0.02	4.0	20	–	–
			0.05	10.0	50	–	–
			0.1	Nil[a]	Nil[a]		
Combined first and second extracts							
			0.02	20.0	100		555
			0.05	20.0	100	–	667
			0.1	Nil	–		

[a] Organic solvent completely dissolved in organic phase.

$$\left(V_w + \frac{S_2 V_w}{100} - \frac{S_1 V_s}{100}\right) - \frac{S_1 V_s}{100} = V_w\left(1 + \frac{S_2}{100}\right) - 2\frac{S_1 V_s}{100} \text{ ml water}$$

$$\text{and } V_s + \frac{S_1 V_s}{100} = V_s\left(1 + \frac{S_1}{100}\right) \text{ ml solvent.}$$

At the end of the second extraction the $Vs\ (1 + S_1/100)$ ml of solvent contains the residual $A\ V_w S_2/100\ V_s\ \mu g$ solute, i.e. at the end of the second extraction the combined organic extracts contain the theoretical initial solute content ($A\ \mu g$) of the original water sample:

$$A\left(1 - \frac{V_w S_2}{100\ V_s}\right) + \frac{A\ V_w S_2}{100\ V_s} = A$$

This data is summarized below:

	Initially	End of first extraction	End of second extraction	Combined organic extracts
Volume of water	V_w	$V_w\left(1+\frac{S_2}{100}\right) - \frac{V_s S_1}{100}$	$V_w\left(1+\frac{S_2}{100}\right) - \frac{2S_1 V_s}{100}$	
Volume of solvent	V_s	$V_s\left(1+\frac{S_1}{100}\right) - \frac{V_w S_2}{100}$	$V_s\left(1+\frac{S_1}{100}\right)$	$2\ V_s\left(1+\frac{S_1}{100}\right) - \frac{V_w S_2}{100}$
Wt of solute in water	A	$\frac{A\ V_w S_2}{100\ V_s}$	Nil	
Wt of solute in solvent	Nil	$A\left(1-\frac{V_w S_2}{100\ V_s}\right)$	$\frac{A\ V_w S_2}{100\ V_s}$	A

The preconcentration factor, i.e. $\frac{\text{volume of water sample}}{\text{volume of total organic extracts}} = \frac{V_w}{V_s(1 + S_1/100) - V_w S_2/100}$

(single extraction) or $\frac{V_w}{2V_s(1 + S_1/100) - V_w S_2/100}$ (two extractions).

Some theoretical results are illustrated in Table 6. If 1 litre water is extracted with solvent then, depending on the solubility of solvent in the aqueous phase, some 50–95 per cent of solute in the aqueous phase is extracted in a single solvent extract and 100 per cent in two solvent extractions. Concentration factors achieved are between 100 and 200 using a single extraction approximately halving to 51–66 when using a double extraction. Appreciably higher preconcentration factors of 1996 (single extraction) or 666 (double extraction) are obtained when 1 litre of water is extracted with 1 ml of solute, 50 per cent of the solute is obtained in a single extract and 100 per cent in a double extraction. When the

Table 7 Interaction of preconcentration factor and detection limit.

Volume of water	Volume of solvent (ml)	Preconcentration factor	If 1 $\mu g\, l^{-1}$ can be detected in the aqueous phase then the following concentration can be determined by preconcentration
1 000	10	105–200	0.009 5–0.005
	2 × 10	51–66	0.019 6–0.015 1
1 000	1	1 996	0.000 50
	2 × 1	666	0.001 51
10 000	10	1 250–2 000	0.008 0–0.000 5
	2 × 10	555–667	0.001 80–0.001 49

volume of water taken is increased to 10 litres and two 10-ml solvent extracts are prepared, depending on the solubility of the solvent in the aqueous phase, 50–80 per cent of solute is recovered in a single solvent extract and 100 per cent in two solvent extracts. Preconcentration factors are between 1250 and 2000 (single solvent extraction) and 555 and 667 (double solvent extraction).

The highest preconcentration factors of all are obtained in those circumstances where the volume of solvent used is small relative to the volume of water and where the solubility of the solvent in the aqueous phase is lowest. Multiple solvent extractions, due to the larger solvent volume always reduce preconcentration factors, with consequent adverse effects on detection limits (Table 7). Consequently, in devising such procedures compromises have often to be reached.

1.9.2 Theory of partition coefficient effects

When an organic Solvent is shaken with an aqueous phase containing a solute the solute distributes itself between the aqueous and organic phases. If a volume of water containing A_1 μg solute is shaken with a volume of organic solvent, then at the end of the extraction W_s μg of solute is present in the organic phase and W_w μg remains in the aqueous phase. This partitioning of the solute at a fixed ratio of water to solvent volumes can be represented as a partition coefficient K as follows.

$$K = \frac{\text{weight of solute in solvent } (A_1 - W_w\, \mu g)}{\text{weight of solute in aqueous phase } (W_w\, \mu g)}$$

The higher W_s $(A_1 - W_w)$ is relative to W_w, i.e. the higher K, the better the extraction from the aqueous to the organic phase and the better the preconcentration achieved. At the end of the extraction we are left with aqueous phase still containing W_w μg unextracted solute. We have at this stage

$$W_{w_1} + W_{s_1} = A_1 \quad \text{and } K = \frac{W_{s_1}}{W_{w_1}}$$

(the suffixes 1 indicating the extraction number)
Putting in terms of W_{w_1}

$$\text{as} \quad W_{s_1} = KW_{w_1}$$

$$W_{w_1} + KW_{w_1} = A_1, \text{ i.e. } W_{w_1}(1 + K) = A_1$$

$$\therefore W_{w_1} = \frac{A_1}{1 + K}$$

Putting in terms of W_{s_1}

$$\text{as} \quad W_{w_1} = \frac{W_{s_1}}{K}$$

$$\frac{W_{s_1}}{K} + W_{s_1} = A_1, \text{ i.e. } W_{s_1}\left(\frac{1}{K} + 1\right) = A_1$$

$$\therefore W_{s_1} = \frac{A_1}{(1 + K)/K} = \frac{A_1 K}{1 + K}$$

Similarly, when we come to perform a second extraction of the residual water phase with a further equal volume of fresh organic solvent the $A_1/(1 + K)$ $(= W_{w_1})$ μg solute in the aqueous phase redistributes itself between the organic and aqueous phases. If A_2 μg represents the total weight of solute available for redistribution (where the suffix 2 indicates the second extraction)

$$\text{as} \quad K = \frac{W_{s_2}}{W_{w_2}} \qquad W_{s_2} = KW_{w_2}$$

$$\text{and} \quad A_2 = K\,W_{w_2} + W_{w_2} = \frac{A_1}{1 + K}$$

$$\text{it follows that} \qquad W_{w_2}(1 + K) = \frac{A_1}{1 + K}$$

$$\therefore W_{w_2} = \frac{A_1}{(1 + K)^2}$$

$$\text{also,} \qquad W_{w_2} = \frac{W_{s_2}}{K}$$

$$\therefore A_2 = W_{s_2} + \frac{W_{s2}}{K} = \frac{A_1}{1+K}$$

$$\therefore W_{s_2}\left(1 + \frac{1}{K}\right) = \frac{A_1}{1+K}$$

i.e. $$W_{s_2} = \frac{A_1}{1 + K/(1 + 1/K)} = \frac{A_1 K}{(1+K)^2}$$

Summarizing:

	μg solute in aqueous phase	μg solute in organic phase
First extraction:	$W_{w_1} = \frac{A_1}{1+K}$	$W_{s_1} = \frac{A_1 K}{1+K}$
Second extraction:	$W_{w_2} = \frac{A_1}{(1+K)^2}$	$W_{s_2} = \frac{A_1 K}{(1+K)^2}$
Similarly, for a third extraction:	$W_{w_3} = \frac{A_1}{(1+K)^3}$	$W_{s_3} = \frac{A_1 K}{(1+K)^3}$

Some results in Table 8 illustrate that a K value of 2 gives only 66.7 per cent recovery of solute in the organic phase in a single extraction, improving to 90 per cent in two extractions and 96 per cent in three extractions. When the K value is 20 then 95 per cent solute recovery is obtained in a single solvent extraction and above 99 per cent in two extractions. For K values above say 100 a single solvent extraction is sufficient to achieve 99 per cent plus solute recovery in the organic phase.

Summarizing, in solvent extraction preconcentrations, highest solute recoveries and preconcentration factors result under the following circumstances:

1. Volume of solvent used small relative to volume of water sample.
2. Choosing a solvent with a low solubility in the aqueous phase.
3. Choosing a solvent with a high K value (where K = wt of solute in solvent/wt of solute in water).

The overall effect of the number of solvent extractions on solute recovery in the organic phase is illustrated in Table 9,[12] which shows recoveries obtained from a range of organics by extracting 1 litre of a water sample with 0.2 ml hexane. Due to solubility effects 0.05 ml of hexane was recovered, i.e. 0.15 ml of hexane had dissolved in 1 litre of water, solubility of hexane in water 0.015 per cent. Both solubility and partition effects are reflected in these results. Both effects are felt in a single extraction where recoveries are between 47.9 and 69.3 per cent, but

Table 8 Influence of partition coefficient on solute recovery.

K	Weight of solute in initial water sample, A_1 (μg)	Weight (μg) of solute in organic extract				Recovery (per cent)			
		1st Ws_1 $\frac{A_1K}{1+K}$	2nd Ws_2 $\frac{A_1K}{(1+K)^2}$	3rd Ws_3 $\frac{A_1K}{(1+K)^3}$	Cumulative Ws_{cum} $\frac{A_1K((1+K)^2+(1+K)+1)}{(1+K)^3}$	1st $\frac{(A_1K)100}{(1+K)A_1}$	2nd $\frac{(A_1K)100}{(1+K)^2A_1}$	3rd $\frac{(A_1K)100}{(1+K)^3A_1}$	Cumulative
2	20	13.33	4.444	1.481	19.25	66.7	22.22	7.41	96.2
20	20	19.05	0.907	0.043	20.00	95.2	4.54	0.22	100.0
200	20	19.90	0.099	0.000 99	20.00	99.5	0.50	0.005	100.0
2000	20	19.99	0.010	0.000 0	20.00	100.0	0.050	0.000	100.0

Table 9 Recoveries of organics in hexane extraction of water.[a]

Organic	Per cent recovery	
	One extraction	Three extractions
Aldrin	47.9	89.4
Heptachlorexpoxide	58.5	91.3
α-Cis-Chlordane	59.2	92.0
Dieldrin	62.2	92.6
n-Decane	69.3	98.6
n-Dodecane	60.3	97.3
n-Tetradecane	56.0	96.7
n-Hexadecane	52.9	96.4
Di-*n*-butyl phthalate	65.5	86.6

[a] 1 litre water extracted with 0.2 ml hexane, 0.05 ml hexane recovered.
From Murray[12] with permission.

only the partition effect has any influence when three solvent extractions were performed with recoveries in the range 86.6 to 98.6 per cent.

Having obtained the organic extract some workers in an attempt to further improve preconcentration factors evaporate the extract to a smaller volume prior to final chemical analysis. In Table 10 are shown some results obtained[12] by concentrating 10 ml hexane extract to 1 ml

Table 10 Losses of organics in evaporating hexane solutions tenfold.

Organic	Per cent losses in evaporating 10 ml hexane to 1 ml		Combined per cent partition effect loss and per cent losses on evaporating 10 ml hexane to 1 ml[a]	
	Synder evaporator at 90 °C	Rotary evaporation at 50 °C	Synder evaporator at 90 °C	Rotary evaporation at 50 °C
Aldrin	77	84	69	75
Heptachlorexpoxide	78	90	71	74
α-Cischlordune	78	87	72	80
Dieldrin	79	85	73	79
n-Decane	57	25	56	25
n-Dodecane	59	39	57	38
n-Tetradecane	61	49	59	47
Di-*n*-butyl phthalate	84	100	71	87

[a] Using recoveries in Table 9.
From Murray[12] with permission.

by two methods. Clearly such procedures are to be avoided as they introduce further serious losses of organics in addition to those discussed above.

1.10 Preconcentration of organics by head-space analysis

In this technique (discussed in Chapter 6, Volume 2) the aqueous sample is enclosed in a container under a head space of a relatively small volume of inert gas. Volatile organics redistribute themselves between the liquid and the gas phases during an equilibration period and then a portion of the head-space gas is withdrawn for analysis by a suitable technique. Three factors affect the enrichment obtained in the gas phase as expressed by the partition coefficient K. These are the water temperature, the presence of added inorganic salts in the water which have an appreciable effect on K, and the pH of the water sample which has a small but important effect on K. In Table 11 are shown the effects of these factors on the preconcentration of a solution of methyl ethyl ketone in in water.[13] It is apparent from these results that by raising sample temperature to 50 °C and adjusting the sodium sulphate content to 3.35 M for pH 7.1 a 260/3.9 = 67 improvement in preconcentration factor has been very simply achieved.

1.11 Preconcentration of organics by purge and trap analysis

In this procedure (discused in Chapter 6, Volume 2) an inert gas is purged through the water sample, sometimes using a closed loop system, and then the gas is passed through a suitable adsorbent usually activated

Table 11 Effect of sample temperature, added salt content, and pH on partition coefficient between aqueous methyl ethyl ketone and nitrogen.

pH	Temperature (°C)	$k \times 10^{-3}$ for for salt concentration (mol Na_2SO_4)		
		0.6	1.41	3.35
4.5	30	4.19	21.3	118
	50	21.3	39.8	234
7.1	30	3.9	20.0	109
	50	20.0	37.6	260
9.1	30	4.56	18.7	105
	50	18.7	35.0	229

From Friant[13] with permission.

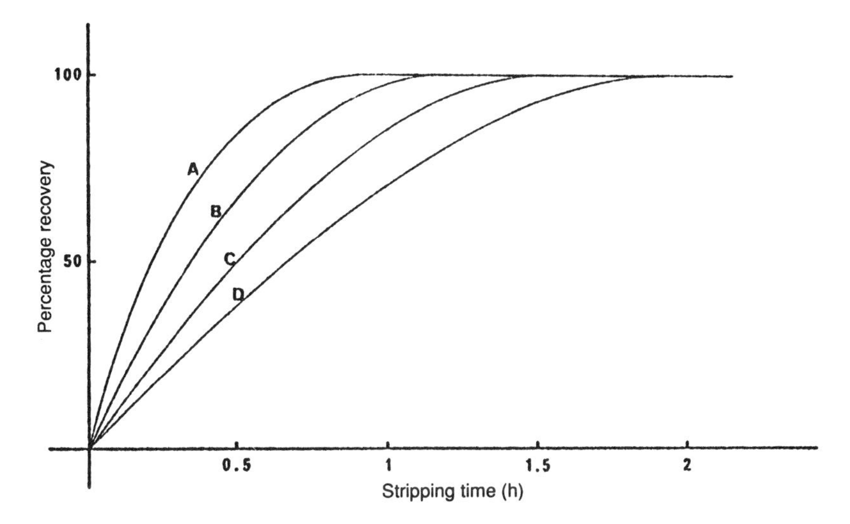

Fig. 10 Percentage recovery of stripped compounds with increasing stripping time: A, *n*-hexane; B, *n*-octane; C, *n*-decane; D, *n*-dodecane. From Colenutt and Thorburn[15] with permission

carbon which retains organics removed from the water sample. The volatiles are then removed from the carbon by a relatively small volume of solvent, or thermally, and analysed by a suitable technique. Several factors affect the efficiency with which carbon retains organics including flow rate of stripping gas, stripping time (in closed loop systems), particle size of carbon, and choice of desorption solvent. The effect of stripping time on recovery is illustrated in Fig. 10[14,15] which shows that stripping times of 1 h suffice for relatively volatile hydrocarbons such as *n*-hexane but that 2 h are required for less volatile *n*-dodecane. Clearly, such studies are always required before the technique can be applied to

Table 12 Effect of particle size of carbon on recovery of organics.

Organic	Per cent recovery for particle size (mm dia.):				
	0.35	0.25	0.18	0.15	0.12
n-Octane	92	95	98	99	99
Toluene	91	95	99	100	100
Ethanol	86	90	93	95	96
Phenol	90	93	95	96	97
Methyl ethyl ketone	86	90	92	94	94
Mean	91	93	95	97	97

From Colenutt and Thorburn[15] with permission.

Table 13 Effect of desorption solvent on efficiency of removing adsorbed organics from carbon.

Adsorbate	Per cent recovery with adsorption solvent:					
	n-Hexane	Toluene	CS_2	Ether	Carbon tetrachloride	Methanol
n-Octane	99	97	99	97	98	90
Toluene	97	–	100	97	96	89
Ethanol	90	90	95	92	88	95
Phenol	94	91	96	92	90	96
Methyl ethyl ketone	91	90	94	93	91	92

From Colenutt and Thorburn[15] with permission.

any particular organic. Table 12 illustrates the effect of carbon particle size on the recovery of various organics showing that smaller particle size carbons are more effective.

Table 13 shows the effect of the desorption solvent chosen on the efficiency of recovery of various adsorbed organics from carbon. Clearly hydrocarbons (*n*-hexane and toluene), carbon disulphide, ether, and carbon tetrachloride are more efficient in desorbing hydrocarbons (*n*-octane and toluene) than they are in desorbing polar compounds such as ethanol, phenol, and methyl ethyl ketone. The situation is reversed with methanol desorbent which desorbs polar compounds more efficiently than hydrocarbons.

1.12 Cross referencing of determinants and preconcentration methods

The breakdown of the material into chapters discussing types of technique with subsections on types of substances, is preferred to a layout based on chapters dealing with types of substances and subsections dealing with type of preconcentration techniques as the former layout enables the subject to be discussed in a more logical and developed manner. Those who are interested in ascertaining what preconcentration techniques are available for particular substances should refer to Table 14.

References

1. Subramanian, K. S. and Meranger, J. C. (1979). *International Journal of Environmental Analytical Chemistry*, 7, 25.
2. Tessier, A., Campbell, P. G. C., and Bisson, M. (1979). *International Journal of Environmental Analytical Chemistry*, 7, 41.

Table 14 Cross referencing of determinands and preconcentration method.

Type of determinand	Volume 1: Chelation–solvent extraction Non-saline	Sea	Adsorption on non-polar resins Non-saline	Sea	Cation exchange resins Non-saline	Sea	Anion exchange resins Non-saline	Sea	Volume 2: Solid adsorbents Non-saline	Sea	Coprecipitation methods Non-saline	Sea	Solvent extraction Non-saline	Sea	Purge and trap head-space analysis Non-saline	Sea	Miscellaneous techniques[a] Non-saline	Sea
Cations	2.1	2.2	3.2	3.3	4.1.1, 4.1.3, 4.1.4	4.1.2, 4.1.3	–	–	1.1.1–1.1.3, 1.1.4–1.1.8, 2.1, 3.1.1.	1.1.1, 1.1.2, 1.1.4, 1.1.5 2.1.1–2.1.3, 2.1.5, 3.1,2	4.2.1, 4.2.2, 4.2.4–4.2.8, 4.6	4.2.1–4.2.3, 4.2.6–4.2.8, 4.6.1, 4.6.4, 4.6.5	–	–	–	–	A:7.1.1 B:7.1.2 C:7.1.3 D:7.1.4 E:7.1.5 F:7.1.6	A:7.1.1 B:7.1.2 C:7.1.3 D:7.1.4
Anions	2.3	2.4	–	–	–	–	5.1, 5.2.1	5.2.2	1.2, 2.2, 3.2	–	4.3	–	–	–	–	–	c:7.2	–
Organics	–	2.5	3.1.1, 3.1.2	3.1.1	4.2	4.2	5.3	–	1.3, 2.3, 3.3.1	1.3.1, 1.3.5, 2.3.1, 3.3.2	–	4.4	5.1	5.2	6.1, 6.2.1	6.1, 6.2.2	C:7.3.2 D:7.3.3 E:7.3.4 F:7.3.5	A:7.3.1 C:7.3.2 D:7.3.3 F:7.3.5
Organometallics	2.6	2.7	–	3.3	4.3	4.3	5.4	–	1.4, 2.4	–	–	–	5.3	5.3.1	–	–	A:7.4	–

[a] Miscellaneous methods A, electrolytic methods; B, gas evolution methods; C, flotation methods; D, evaporation/distillation/freeze drying methods; E, precipitation/cocrystallization methods; F, dialysis/osmosis methods.

3. Reggeks, G. and Van Grieken, R. (1984). *Fresenius Zeitschrift Fur Analytische Chemie*, *317*, 520.
4. Burba, P. and Willman, P. G. (1983). *Talanta*, *30*, 381.
5. Tateda, A. and Fritz, J. S. (1978). *Journal of Chromatography*, *152*, 329.
6. Ryan, J. P. and Fritz, J. S. (1978). *Journal of Chromatographic Science*, *16*, 448.
7. Matthews, K. M. (1972). *Analytical Letters*, *16*, 633.
8. Glueckauf, A. (1983). *Ion Exchange and its Analytical Applications*. Society of Chemical Industry, London (1955).
9. Mayer, A. and Tomkins, C. (1947). *Journal of the American Chemical Society*, *69*, 2866.
10. Glueckauf, A. (1955). *Transactions of Faraday Society*, (London) *51*, 34.
11. Chakrovarty, R. and Van Grieken, R. (1982). *International Journal of Environmental Analytical Chemistry*, *11*, 67.
12. Murray, D. A. J. (1979). *Journal of Oceanography*, *177*, 135.
13. Friant, S. L. and Suffet, T. (1979). *Analytical Chemistry*, *51*, 2167.
14. Colenutt, B. A. and Thorburn, S. (1980). *International Journal of Environmental Analytical Chemistry*, *7*, 23.
15. Colenutt, B. A. and Thorburn, S. (1980). *International Journal of Environmental Studies*, *15*, 25.

2

CHELATION-SOLVENT EXTRACTION OF ANIONS AND CATIONS

2.1 Metal cations – non-saline waters

The low concentrations at which metals can occur in certain types of water samples, for example, potable waters, rain water, snow, ice, and sea-water, preclude their direct determination by even the most recent advanced methods of analysis. To overcome this problem and improve the effective detection limits of these techniques various methods have been devised for the preconcentration of samples prior to analysis.

One such method is complexation–solvent extraction wherein the metals in a large volume of water sample are reacted with an organic complexing agent dissolved in a small volume of an organic solvent. The solvent is then either analysed by direct aspiration into the atomic absorption spectrophotometer, or other suitable instrument or it is back-extracted with a small volume of aqueous acid which is then analysed. Either way, a preconcentration factor is achieved which is approximately equal to the ratio of the original volume of water sample taken to the volume of the final extract analysed. The detection limit of the preconcentration method relative to the original unmodified method will be improved by approximately this ratio.

The considerable difficulty of trace element analysis in a high salt matrix such as sea-water, estuarine water or brine is clearly reflected in the literature. The extremely high concentrations of the alkali metals, the alkaline earth metals, and the halogens, combined with the extremely low levels of the transition metals and preconcentration procedures based on chelation and solvent extraction assist here by separating the metals to be determined from the sample matrix, and, additionally, serve to lower detection limits to an acceptable level.

An ideal method for the preconcentration of trace metals from natural waters should have the following characteristics: it should simultaneously allow isolation of the analyte from the matrix and yield an appropriate enrichment factor: it should be a simple process, requiring the introduction of few reagents in order to minimize contamination, hence producing a low sample blank and a correspondingly lower detection limit; and it should produce a final solution that is readily matrix matched with solutions of the analytical calibration method. Various organic chelating agents have been studied, these are discussed below.

2.1.1 Dithiocarbamic acid derivatives

Atomic absorption spectrometry Kinrade and Van Loon[1] as early as 1974 pointed out that although solvent extraction methods for concentrating traces of metal ions in water abound, very little work had been done up to that time on the optimization of experimental conditions making it impossible to obtain good results on a routine basis. These workers carried out a systematic examination of the factors which have a bearing on the quality of results including pH dependence, variation in the ratio of the two complexing agents used (ammonium pyrrolidinedithiocarbamate and diethylammonium diethyldithiocarbamate dissolved in methyl isobutyl ketone), extraction time, and reagent stability. They showed that eight metals, cadmium, cobalt, copper, iron, lead nickel, silver, and zinc, could be simultaneously extracted by their finally evolved method with good sensitivity and precision (Tables 15 and 16).

Subramanian and Meranger[2] made a critical study of the solution conditions and other factors affecting the reliability of the ammonium pyrrolidinedithiocarbamate–methyl isobutyl ketone extraction system for the determination of silver, cadmium, cobalt chromium, copper, iron manganese, nickel, and lead in potable water. Graphite furnace atomic absorption spectrometry was used for the finish. The following parameters were investigated in detail: pH of the aqueous phase prior to

Table 15 Precision measurements expressed as per cent coefficient of variation.

Solution concentration (μg litre^{-1})	Ag	Cd	Co	Cu	Fe	Ni	Pb	Zn
250 (solutions shaken)	7.16	0.26	0.67	0.52	0.99	0.37	0.41	0.29
25 (solutions inverted)	1.86	0.92	5.23	4.27	6.46	5.51	4.26	2.25

From Kinrade and Van Loon[1] with permission.

Table 16 Sensivity and range of linearity.

	Ag	Cd	Co	Cu	Fe	Ni	Pb	Zn
Sensitivity (μg litre^{-1})	0.6	0.8	1.5	0.8	1.3	1.3	2.5	0.6
Range of linearity (μg litre^{-1})	0–200	0–400	0–350	0–400	0–350	0–300	0–5000	0–200

From Kinrade and Van Loon[1] with permission.

extraction; amount of ammonium pyrrolidinediethyldithiocarbamate added to the solution following pH adjustment; the length of time needed for complete extraction; and the time stability of the chelate in the organic phase. Except for silver and chromium which were quantitively extracted only in a very narrow pH range (1.0–2.0 and 1.8–3.0, respectively) and cadmium and lead which were stable in the extracted methyl isobutyl ketone phase only for 2–3 h, the solution conditions for quantitative extraction were not critical for the other metals. Simultaneous extraction of all the metals except cadmium and lead was also investigated. Good recoveries (100 ± 10 per cent) were obtained for a number of spiked raw treated and distributed potable water samples covering a wide range of hardness. They concluded that the procedure is reliable and precise under proper solution conditions.

The optimized conditions are embodied in the following method which is quoted as an example of experimental technique. One important feature of preconcentration methods is featured in the method, namely the purification of the aqueous methyldithiocarbamate reagent and the buffer solution by extraction with a methyl isobutyl ketone solution of the chelating agent.

Apparatus:

Perkin Elmer Model 603 atomic absorption spectrophotometer equipped with a HGA-2100 graphite furnace and a deuterium arc background corrector or equivalent. Single element hollow cathode lamps used as narrow line sources for all the elements except cadmium and lead which were determined by electrodeless discharge lamps to enhance sensitivity. Nitrogen purge gas used for all the elements except chromium for which argon was used. Cadmium and lead were determined in the gas interrupt mode.

Nalgene (conventional polyethylene) 1000 ml screw cap bottles (containers for water samples).

125 ml Pyrex glass separatory funnels fitted with Teflon stopcocks and polyethylene stoppers.

Reagents:

High purity water: distilled doubly deionized water in a Corning all glass distillation system.

Standard solutions of metals in 0.1 per cent nitric acid (Baker Ultrex): prepare by serial dilution of the 1000 mg litre^{-1} stock solutions.

Ammonium citrate buffer (20 per cent w/v): prepare by dissolving 200 g of ammonium citrate, dibasic ACS grade, Fisher Scientific, in 500 ml water and adjust the pH to 7.2 with concentrated ammonia. Make up the solution to 1 litre with water and extract for 3 min with 5 ml of a 1 per cent solution of purified ammonium pyrrolidinediethyldithilocarbamate and 25 ml methyl isobutyl ketone. Repeat this operation until the aqueous layer is virtually free of any trace metal impurities. Store the aqueous phase in a 1-litre pre-cleaned polyethylene bottle.

Ammonium pyrollidine diethyl diethiocarbamate: prepare by dissolving 20 g of the compound (Baker analysed) in 1 litre of water and extract for 3 min with 50 ml methyl isobutyl ketone. Discard the ketone layer and repeat the extraction until the organic phase becomes colourless. Store the aqueous phase, in a pre-cleaned polyethylene bottle at room temperature. This solution is stable for at least 1 month.

Methyl isobutyl ketone (or 4-methyl-2-pentanone) supplied by Fisher Scientific (ACS grade). Use without further purification.

Procedure:

Determination of cadmium and lead: Transfer samples of raw, treated or distributed water (25 ml) to a 125 ml separatory funnel. Add ammonium citrate buffer (10 ml) and adjust the pH to 4.6 then add 5 ml of 2 per cent ammonium pyrrolidinedithiocarbamate and 5 ml of methyl isobutyl ketone. Extract the solution for 2 min prior to the determination of cadmium by heated graphite atomization. In the case of lead analysis, add 5 ml of 0.05 per cent ammonium pyrrolidinedithiocarbamate prior to extraction.

Determinations of silver, cobalt, chromium, copper, iron, manganese, and nickel: Use the same method as above except for the pH which should be adjusted to 2.0–2.5. Do not add a buffer.

Determine the metal concentration by injecting 20 μl of the methyl isobutyl ketone phase into the graphite furnace using an 'Eppendorf' pipette fitted with disposable polypropylene tips. Decontaminate the tips from traces of metals by soaking them overnight in 1 per cent nitric acid (Baker Ultrex). Follow the optimum 'dry', 'Char', and 'atomize' programme of HGA-2100 in Table 17. Note the peak absorbance.

In Figs 1 to 3 is shown the effect of sample pH on the extraction efficiency of several metals. Figure 4 shows the effect of complexing agent: metal ratio on extraction efficiency. Metal recoveries are satisfactory as shown in Table 18.

Bone and Hibbert[3] have described a method using ammonium pyrrolidinedithiocarbamate dissolved in 2,6-dimethyl-4-heptanone followed by atomic absorption spectrometry for the determination of vanadium, chromium, iron, cobalt, nickel, copper, zinc, molybdenum, cadmium, and lead in effluents and natural waters.

Table 17 Optimized instrumental parameters for trace metals in the methyl isobutyl ketone phase using graphite furnace atomic absorption spectrometry.

Element	Line (nm)	Slit (nm)	Ash		Atom	
			Temp. (° C)[a]	Time (s)	Temp. (° C)[a]	Time (s)
Ag	328.1	0.7	300	10	2 400	8
Cd	228.8	0.7	by-pass		1 600	7
Co	240.7	0.2	500	10	2 600	10
Cr	357.9	0.7	500	10	2 600	10
Cu	324.8	0.7	600	10	2 500	7
Fe	248.3	0.2	600	10	2 600	8
Mn	279.5	0.2	500	10	2 600	7
Ni	232.0	0.2	500	10	2 600	9
Pb	283.3	0.7	400	10	2 200	7

From Subramanian and Meranger.[2]
[a]Drying temperatures: Cd, 100 °C for 50 5; other metals, 100 °C from 305.

Procedure:

Oxidation of samples: Transfer duplicate 300 ml volumes of each sample to 500 ml conical flasks. Place a watch glass over each flask, add 5 ml aliquot of 1 per cent wlv cerium (IV) in 2 per cent wlv sulphuric acid to each solution, and heat the flasks on a steam bath for 20 min. If, during heating, the yellow colour of cerium(IV) was discharged, add more oxidant. After digestion allow the samples to cool and transfer to 500 ml separating funnels.

Extraction procedure: Add sufficient 50 per cent wlv ammonium acetate buffer to adjust the pH to the required value in the range 3–6 (to avoid contamination, determine the amount of buffer needed to achieve the desired pH on a separate aliquot of the sample). After the addition of 20 ml of the chelate solution and 8 ml of di-isobutyl ketone, shake the separating funnels by hand for 2 min. After at least 5 min for phase separation, discard the aqueous phase or collect it for pH measurement. Collect the organic phases in dry, stoppered flasks and subsequently analyse by flame AAS. Determine reagent blanks by using deionized water purified with dithizone and the same batch of 50 per cent (v/v) nitric acid that was used during sampling. Achieve calibration of the blank by adding working standard to the purified water. Establish the zero setting by aspirating pure ketone into the flame. Aspirate deionized water and 6 per cent (v/v) nitric acid periodically to prevent nebulizer clogging.

Table 18 Per cent recovery of metals in spiked samples of raw, treated, and distributed potable water using APDC-MIBK-GFAA.

Concentration of spike ($ng\ mL^{-1}$)	Recovery (per cent)								
	Ag	Cd	Co	Cr	Cu	Fe	Mu	Ni	Pb
0.2	—	96 ± 7[a]	—	—	—	—	—	—	—
0.4	100 ± 3	101 ± 4	—	—	—	—	94 ± 6	—	—
0.6	104 ± 2	98 ± 3	—	—	—	—	—	—	—
0.8	—	85 ± 4	—	—	—	—	103 ± 5	—	—
2.0	103 ± 3	—	—	92 ± 6	97 ± 4	—	93 ± 4	—	104 ± 5
4.0	100 ± 4	—	94 ± 5	96 ± 3	105 ± 2	98 ± 3	97 ± 3	—	—
6.0	98 ± 3	—	—	—	101 ± 3	—	—	—	95 ± 5
8.0	—	—	96 ± 3	97 ± 2	105 ± 2	—	—	—	98 ± 6
10.0	—	—	—	97 ± 4	104 ± 4	96 ± 1	98 ± 2	—	101 ± 5
20.0	—	—	97 ± 2	91 ± 6	95 ± 2	93 ± 2	—	101 ± 1	—
40.0	—	—	97 ± 4	—	—	102 ± 4	—	104 ± 1	—
50.0	—	—	—	—	—	98 ± 1	—	—	—

[a]Values given represent the average of the triplicate analyses each of 20 raw, treated, and distributed potable water samples ranging in hardness from 1 to 554 mg $CaCo_1$ $litre^{-1}$. The values are more or less the same for single as well as simultaneous extractions. The measure of precision is the standard deviation.

From Subramanian and Meranger[2] with permission.

This method can be recommended for samples containing less than 3 mg of iron. The instability of the manganese chelates made measurement in the organic phase unsatisfactory. However, manganese could be successfully determined in the acid extract of the organic phase. Cerium(IV) sulphate is effective for the oxidation of chromium(III) to chromium(IV) and does not interfere with the extraction. Because of the existence of non-liable species of trace elements, it is advisable to acidify and oxidize samples prior to extraction and cerium(IV) also can be used for this purpose.

Particularly in the analysis of effluents, calibration by the method of standard additions is essential for accurate results. Precisions are between 1.4 per cent (copper) and 6.9 per cent (chromium). Linear ranges are between 0–40 μg litre^{-1} (chromium, iron, cobalt, nickel, copper, zinc) and 0–300 μg litre^{-1} (manganese). The working pH ranges are 3.4–4.6 (manganese), 3.4–5.1 (iron, vanadium, cobalt, nickel, copper, molybdenum and cadmium), 3.4–5.6 (zinc, lead), and 4.1–5.6 (chromium).

Tessier *et al.*[4] evaluated an ammonium pyrrolidinedithiocarbamate methyl butyl ketone preconcentration procedure for the determination of traces of cadmium, cobalt, copper, nickel, lead, zinc and molybdenum in river water samples. These workers reported that sample contamination, analyte ions, and diverse matrix effects can all adversely affect the reliability of methods. They studied the effect of sample composition (i.e. its matrix) on trace metal recovery and re-examined the effect of sample pH on metal recovery. The influence of sample pH on the overall efficiency of the chelation–extraction procedure was determined at intervals of 0.5 pH units over the pH range 1.5–9.

The results, expressed as relative absorbance show that in the same pH range 3–8 the extraction efficiency is virtually independent of pH for six of the seven metals studied; a slight decrease is observed at high pH values for copper, lead, and zinc, whereas a marked reduction in extraction efficiency is noted at pH values less than 3 for all metals but manganese.

The simultaneous extraction of cadmium, cobalt, copper, nickel, lead, and zinc is clearly feasible in the pH range 3 to 8. Tessier *et al.*[4] chose an intermediate sample pH of 4; this corresponds to an extraction pH of approximately 4.9, the shift in pH being due to the addition of ammonium pyrrolidinedithiocarbamate. In the case of molybdenum, however, the marked decrease in relative absorbance in the pH range 2–4 effectively precluded its analysis at a sample pH of 4.

Plots of absorbance as a function of trace metal concentration in the initial aqueous sample showed a linear relationship in the following concentration ranges: copper, nickel and cobalt, 0–100 μg litre^{-1}; lead, 0–30 μg litre v^{-1} cadmium and zinc, 0–20 μg litre^{-1}. The sensitivity of

Table 19 Sensitivity and precision of the chelation/extraction procedure as determined on natural water samples.

Metal	Sensitivity[a] (μg litre^{-1})	Coefficient of variation[b]	
		Filtered sample (%)	Unfiltered sample (%)
Cadmium	0.6	6.7 (3.4)	2.5 (5.8)
Cobalt	2.9	3.3 (5.6)	2.4 (5.2)
Copper	2.1	2.7 (15.3)	5.1 (39.6)
Nickel	2.2	1.8 (48.1)	2.3 (45.5)
Lead	4.4	10.3 (5.9)	3.2 (49.1)
Zinc	0.7	3.4 (58.3)	1.6 (103)

[a]Sensitivity is defined as the concentration needed to produce a 1 per cent absorption.
[b]The coefficient of variation was obtained from 10 replicate analyses; the metal concentration (μg litre^{-1}) is given in parentheses.
From Tessier *et al.*[4] with permission.

the overall procedure, defined as the metal concentration needed in the original sample to obtain a 1 per cent absorption after chelation and extraction, is given in Table 19. Comparison of these values with those obtained on the same instrument by direct aspiration of an aqueous sample shows a 17- to 36-fold increase in sensitivity. The detection limits, defined as the concentrations needed to obtain a signal equal to twice the baseline variation, are cadmium, 0.2 μg litre^{-1}; copper and zinc, 0.5 μg litre^{-1}; nickel, 1.5 μg litre^{-1}; cobalt, 2.0 μg litre^{-1}; lead, 2.5 μg litre^{-1}.

Tessier *et al.*[4] found no correlation between recovery of trace metals in spiked samples obtained by the chelation method and the electrical conductivity or soluble organic contents of the original sample. The absence of correlation with these intergrative parameters does not however rule out possible effects of specific compounds upon metal recovery. Further examination of Table 20 indicates that maximum errors of about 15 per cent would have been incurred in determining the concentration of trace metal if a calibration curve had been used for quantitation instead of the standard addition technique. From a practical standpoint, in view of the numerous possibilities of error linked to each step from sampling to analysis, the use of a calibration curve prepared in a natural water matrix would seem to represent an acceptable compromise between analytical accuracy and laboratory capacity (i.e. number of samples analyzed). For more accurate results, however, use of a standard addition technique would be required.

Contrary to some early workers, Tessier *et al.*[4] do not find a strong inhibitory effect of linear alkylbenzene sulphonates or humic acid con-

Table 20 Mean values and standard deviation for recovery of spiked metals in the Yamaska and Saint-François river samples relative to that from deionized water.

Campaign	Sample treatment	Cd	Cu	Pb	Zn
1	NF	99 ± 4	99 ± 5	98 ± 7	106 ± 12
	F	100 ± 4	96 ± 6	99 ± 19	104 ± 14
2	NF	109 ± 8	101 ± 6	114 ± 14	101 ± 7
	F	112 ± 5	102 ± 4	119 ± 18	106 ± 11
3	NF	100 ± 4	111 ± 5	92 ± 8	96 ± 9
	F	102 ± 2	104 ± 5	90 ± 13	101 ± 5
4	NF	103 ± 3	104 ± 7	95 ± 11	104 ± 6
	F	106 ± 6	101 ± 8	91 ± 10	108 ± 2

From Tessier *et al.*[4] with permission.

centrations in the original river water sample on trace metal recovery by their chelation–solvent extraction procedure. Ammonium pyrrolidinedithiocarbamate appears to compete effectively with natural complexing agents such as humic acid for the trace metals.

The conclusions so far reached regarding the use of ammonium pyrrolidinedithiocarbamate must be measured up against the comments made by Smith *et al.*[5] who compared eight different preconcentration techniques for the determination of manganese, cobalt, zinc, europium, caesium, and barium in natural waters. These workers used X-ray fluorescence spectrometry as the means of analysing the concentrates. The preconcentration techniques used were: passage through columns of Dowex Al chelating resin and silylated silica gel; filtration through laminate membrane filters and chelating diethylenetriamine cellulose filters; precipitation with sodium diethyldithiocarbamate and 1-(2-pyridylazo)-2-naphthol; extraction with ammonium pyrrolidine dithiocarbamate; and chelation by 8-quinolinol (oxine) followed by adsorption on activated carbon. From the results obtained it seems that chelation by oxine and adsorption on activated carbon gives the highest and most reproducible collection yields for all the waters and elements studied, followed by diethyldithiocarbamate and 1-(2-pyridylazonapthol) precipitation; the results of ammonium pyrrolidinedithiocarbamate extractions were least satisfactory. If one considers the average variation of the collection yields from the different water samples studied, then filtration through silylated silica gel and oxine chelation followed by activated carbon adsorption are least influenced by the water characteristics, while ammonium pyrrolidinedithiocarbamate extraction and diethyldithiocarbamate precipitation suffer most. On average, it appeared that zinc is most easily collected, followed by cobalt, europium, and manganese.

Rubio *et al.*[6] carried out a comparative study of the determination of cadmium, copper, and lead in river water by atomic absorption spectrometry of ammonium pyrrolidinedithiocarbamate–ethyl isobutyl ketone preconcentrates and by direct inductively coupled plasma atomic emission spectrometry. Both methods gave similar detection limits but the ICP method had several advantages including better long term precision and minimal sample preparation. Although similar values for copper, lead, and cadmium were obtained by the two techniques, they are unfortunately, too near to the detection limits for any firm conclusion to be made regarding agreement between the methods.

A mixture of ammonium pyrrolidinedithiocarbamate and diammonium diethyldithiocarbamate in 4-methyl-pentan-2-one has been used as a means of preconcentrating cadmium, zinc, copper, iron, lead, and nickel from sewage effluents.[7] Down to 1 μg litre^{-1} of these elements can be determined satisfactorily in sewage samples using the following method.

Apparatus:

Use all glass apparatus which has been cleaned by boiling with 1 per cent v/v nitric acid. Carry out adequate rinsing with deionized water before use. It is advisable to set aside apparatus specifically for metals analysis, in order to minimize contamination.

Reagents:

Nitric acid SG 1.42 Aristar grade.

4-methylpentan-2-one.

Ammonium tetramethylenedithiocarbamate.

Diethyldiammonium dithiocarbamate.

Sodium citrate, Analar.

Citric acid, Analar.

Ammonia solution SG 0.88.

Chelating solution: 1 per cent w/v each of ammonium pyrrolidinedithiocarbamate and diammonium diethyldithiocarbamate in water.

Buffer solution: 1.2 M sodium nitrate and 0.7 M citric acid. Prior to use, purify the citrate buffer by extracting, using 50 ml of 4-methylpentan-2-one and approximately 0.5 g of ammonium pyrrolidinedithiocarbamate–diammonium diethyldithiocarbamate until the extracts are no longer green (this is usually achieved by two extracts).

Procedure:

Take 500 ml of effluent, add 5 ml nitric acid and boil down to approximately 50 ml in a Kjeldahl flask. When cool make up to the original volume. Take 250 ml of the treated sample and pour into a separating funnel. Neutralize to pH 7 by cautious dropwise addition of ammonia solution (approximately 2.5 ml). Add 5 ml of citrate buffer and check that the pH is around 5.0. Add 10 ml of chelating solution. Add 35 ml of 4-methylpentan-2-one. Shake the funnel for 1 min and then allow to stand for 5–10 min to facilitate separation of the two phases. Discard the aqueous phase and pass the organic phase through phase-separating paper (Whatman IPS) into a 50 ml solumetric flask. Make up to 50 ml with 4-methylpentan-2-one. The solution is now ready for atomic absorption spectrometry.

Allen *et al.*[8] compared methods involving solvent extraction using ammonium pyrrolidinedithiocarbamate–isobutyl methyl ketone with column chelation procedures using immobilized 8-hydroxyquinoline on a controlled glass pore support for the determination of lead and copper in sea and river water. The final determination of the preconcentrated element was accomplished by atomic absorption spectrophotometry using a flame source. Results at the μg^{-1} $litre^{-1}$ level for standard solutions of copper gave recoveries of better than 98 per cent from both procedures. The determination of copper in natural waters showed higher results by the column procedure, suggesting that column extraction was more efficient than solvent extraction. The column procedure was less time consuming and less costly than solvent extraction.

The determination of lead at the $\mu g\ litre^{-1}$ level with copper present gave a recovery of better than 99 per cent when employing 8-hydroxyquinoline column separation. Copper, however, was only separated to the extent of 70 per cent from the same solution of mixed elements.

Reagents:

All reagents used were of Anal R grade. Prepare solutions using distilled deionized water and store in polythene bottles.

Stock copper solution: 100 mg $litre^{-1}$. Prepare from copper(II) sulphate pentahydrate and stabilize with 5.0 ml $litre^{-1}$ of concentrated hydrochloric acid.

Sodium chloride solution, 3 per cent m/v.

Ammonium tetramethylenedithiocarbamate: dissolve 0.25 g pure reagent in 25 ml of distilled, deionized water, and extract the mixture with isobutyl methyl ketone (10 ml). Discard the organic layer.

Hydrochloric acid, concentrated, SG 1.18 Isobutyl methyl ketone. Redistil solvent for extraction.

Apparatus:

Soak all glassware overnight in 25 per cent nitric acid, rinse thoroughly before use with distilled, deionized water, and dry in a dust three atmosphere.

Preparation of samples:

Filter natural water samples through a Whatman No. 1 paper (25 cm) to remove solids, discarding the first 100 ml. Deoxygenate the samples by the passage of oxygen free nitrogen through the bulk sample (20–30 min). If nitric acid was used to stabilize the samples then neutralize them with sodium hydroxide (30 per cent) solution and they acidify to 0.05 M with hydrochloric acid when following the solvent extraction procedure.

Preparation of working standards:

Dilute stock copper solution (100 mg litre^{-1}) to 1 mg litre^{-1} and transfer appropriate volumes via a burette into calibrated flasks (500 ml) to give standard solutions containing 0, 5, 10, 20, and 30 μg litre^{-1} of copper in distilled, deionized water. Prepare the standards freshly for each run of analyses.

Extraction:

Treat samples and standards in the same manner. Acidify the sample (400 ml) in a separating funnel (500 ml) with concentrated hydrochloric acid (1 ml), and shake the funnel. Add ammonium pyrrolidinedithiocarbamate solution (3 ml) and shake the funnel to ensure complete mixing. Add methyl isobutyl ketone (20 ml) and shake the mixture vigorously for 1 min. After settling for a period of 5 min, run off the aqueous layer into a beaker, collect the organic layer in a calibrated flask (25 ml). Extract the aqueous layer once again with methyl isobutyl ketone (5 ml), separate the solvent and combine with the first extract and make up the volume to 25 ml with solvent. Analyse standards and sample extracts by atomic absorption spectrophotometry as soon as possible.

The results obtained with synthetic solutions of copper(II) following ammonium pyrrolidinedithiocarbamate solvent extraction are given in Table 21. In the absence of a 'pure' and stable organocopper standard the calibration solutions were treated in the same manner as was the test samples. Any inaccuracy in the extraction procedure was therefore cancelled out and empirical rather than absolute values were obtained.

It was found that the methyl isobutyl ketone solutions of the copper ammonium pyrrolidinedithiocarbamate extracts were stable for up to 2 h and after that time a noticeable drop in the copper values was observed. The blank value obtained was less than 1 μg litre^{-1} of copper(II) and hardly measurable on the chart recorder plot. Any such blank was taken into account in calculating the percentage recovery.

The results for a river water sample following the two extraction pro-

Table 21 APDC extraction of synthetic samples.

Day	Sample concentration (μg litre^{-1})	Concentration found (μg litre^{-1})	Recovery (per cent)[a]
1st	30	28.5	95.00
	30	27.0	90.00
	30	30.0	100.00
	30	30.5	101.66
	30	30.0	100.00
	30	31.5	105.00
2nd	30	30.0	100.00
	30	30.0	100.00
	30	30.0	100.00
	30	28.0	93.33
	30	30.0	100.00

[a] x = 98.63 per cent; s = 2.56 per cent.
From Allen *et al.*[8]

Table 22 Cu–ammonium and Cu–CPG-8-HOQ column extraction of river water sample 1.

Ammonium pyrrolidinedithiocarbamate		Column CPG-8-HOQ	
Day	Concentration found (p.p.b.)[a]	Day	Concentration found (p.p.b.)[b]
1st	2.25	1st	2.25
	1.75		2.25
	1.75		2.5
	2.25		2.75
	2.25		2.75
			2.75
2nd	1.5		3.25
	2.05		
	2.05	2nd	3.0
	1.5		2.88
			3.12
3rd	1.75		
	2.0	3rd	3.0
	2.25		3.5
	2.25		3.5
	2.25		3.5

[a] x = 1.98 p.p.b.; s = 0.43 p.p.b.
[b] x = 2.93 p.p.b.; s = 0.56 p.p.b.
From Allen *et al.*[8]

cedures are summarized in Table 22. Each set of analyses were carried out on different days. The blank values were minimal and taken into account. Ammonium pyrrolidinedithiocarbamate extraction yielded a mean value of 1.98 μg litre^{-1} of copper(II) and a standard deviation of 0.433 μg litre^{-1} compared with a mean value of 2.93 μg litre^{-1} and a standard deviation of 0.564 μg litre^{-1} for column extraction procedure with respect to the river water sample.

Correct pH conditions are essential for a satisfactory extraction of metal ions in both separation procedures. In the solvent extraction ammonium pyrrolidinedithiocarbamate metal complexes are extractable and stable between pH 1.0 and 5.0 with some selectivity according to the metal ion in question. A controlling buffer or an acid may be employed to adjust the pH of the sample. Hydrochloric acid was found to be satisfactory for pH adjustment. The chelation of metal ions with 8-hydroxyquinoline is pH dependent. The extraction efficiency is also dependent on the flow rate of the sample in contact with it.

Copper ion and lead ion in their pure solutions were satisfactorily extracted at pH 4.6 at flow rates of 5 ml min^{-1}, but in admixture the extraction pattern changed. At pH 2.0 the uptake of lead was very small, but lead was completely extracted in the presence of copper at pH 2.0, at which value only about 70 per cent of the copper was extracted in the presence of lead. At pH 4.6, while lead in the mixture was still 99 per cent retained, the copper uptake had decreased to about 15 per cent, presumably at the expense of lead uptake. A possible explanation of this discrepancy is that there is competition between lead and copper ions, and possibly hydrogen ions, for the 8-hydroxyquinoline sites of the column material. Reducing the flow rate might have increased the copper uptake but one remedy would be to pass the extracted solution (now lead free) through the column a second time to extract the copper ion remaining in the sample.

An application of hexamethyleneammonium hexamethylenedithiocarbamate as a complexing agent involves the extraction of nanogram amounts of cadmium, silver, bismuth, cobalt, copper, nickel, lead, thallium, and zinc in potable water samples.[9] Metals are determined in the solvent extract (xylene–di-isopropyl ketone) by atomic absorption spectrometry or graphite furnace atomic absorption spectrometry.

Apparatus:

Perkin Elmer atomic absorption spectrometer, Model 420 or equivalent with a three slot burner for the air–acetylene mixture and a deuterium back ground compensator.

Perkin Elmer graphite furnace, Model HGA 500.

Reagents:

The inorganic reagents used were of the highest available purity; organic solvents were distilled before use.

Methyl isobutyl ketone.

Di-isopropyl ketone.

Xylene—this solvent is a mixture of the various isomers.

Hexamethyleneammonium hexamethylenedithiocarbamate: to a solution of 224 ml of distilled hexamethyleneimine (boiling point 136–138 °C) in 300 ml of xylene, which is being cooled in an ice-bath, add within 30 min, and with constant stirring and cooling, 60 ml of distilled carbon disulphide (boiling point 46.2 °C). Collect the white crystalline precipitate on a funnel, wash it three times with diethyl ether and then dry it between filter papers.

Formate buffer solution: dissolve 268 g of formic acid and 14 g of citric acid monohydrate in about 350 ml of water. Add slowly, with constant cooling and stirring, 243 g of sodium hydroxide. To this mixture add 50 mg of *m*-cresol purple and dilute the solution to 1 litre with water. Wash this solution twice with 50 ml of extraction solution in order to remove trace amounts of extractable metals.

Extraction solution, 0.025 M hexamethyleneammonium hexamethylenedithiocarbamate: dissolve, in a dry, 250 ml calibrated flask, 1.7 g of hexamethyleneammonium hexamethylenedithiocarbamate in 75 ml of xylene, heating gently if necessary. Adjust the solution to the mark with 2,4-di-isopropyl ketone and keep it in a cool dark place.

Hexamethyleneammonium hexamethylenedithiocarbamate solution, 0.2 M in methanol: in a dry, 100 ml calibrated flask, dissolve 5.5 g of hexamethyleneammonium hexamethylenedithiocarbamate in methanol, heating gently if necessary. Cool the solution to room temperature and adjust it to the mark with methanol.

Aqueous metal standard solutions: prepare aqueous standard solutions by dilution of an aqueous stock solution with 0.1 M nitric acid.

Organic stock solution. This solution contains 50 mg litre^{-1} of silver, bismuth, cadmium, cobalt, copper, nickel, lead, thallium, and zinc in organic solvent. Pipette 5 ml of aqueous stock solution into a dry 100 ml calbrated flask. Add 50 ml of formic acid and 0.25 g of nitric acid monohydrate. Adjust to the mark with methyl isopropyl ketone. Organic metal standard solutions should be kept in a cool, dark place.

Working procedure:

Treat a 400 ml sample of water with 20 ml of extraction solution, corresponding to an enrichment factor of 20. Other volumes, up to a ratio of the volume of the aqueous phase to the volume of the organic phase of 50:1 can be used. Measure 400 ml of the water sample in a graduated cylinder and transfer into a 500 ml calibrated flask. Add 20 ml of formate buffer solution; the colour of the indicator should be a pure yellow. If a red colour appears, add an additional 20 ml of formate buffer solution. Next add 2 ml of hexamethyleneammonium hexamethylenedithiocarbamate solution in methanol and shake the flask vigorously. Wait for 5 min, then add 20 ml of extraction solution and again shake the flask vigorously for at least 3 min.

Wait for about 10 min in order to allow the layers to separate. Then carefully add water until the organic layer is completely in the neck of the flask. Adjust the absorbance reading of the atomic absorption spectrometer to zero while aspirating the extraction, solution. Next, aspirate the organic layers of each prepared sample, the organic layer of a blank treated in the same way as the samples and at least three organic standard solutions per element. In addition, aspirate mixed solvent (30 per cent v/v xylene and 70 per cent v/v di-isopropyl ketone) between each sample and between each standard in order to prevent clogging of the nebulizer. Finally, subtract the blank reading from the observed absorbance of each sample to obtain the true absorbance value of the sample.

Calculation:

Determine the concentration of each metal in the extract of each sample from plots, on linear graph paper, of the absorbances of standards in the respective metal. Calculate the metal concentration of the samples by dividing the concentration values of the sample extracts by the enrichment factor of the extraction.

Table 23 Extraction of cadmium.

Aqueous solution		Extraction solution volume (ml)	Ratio of phases	Recovery per cent	Measurement technique[a]
Concentration (μg litre^{-1})	Volume (ml)				
0.1	20	1	20:1	>95	Graphite furnace
0.2	20	1	20:1	>95	Graphite furnace
0.5	800	16	50:1	>95	Flame
1	800	16	50:1	>95	Flame
2.5	400	20	20:1	>95	Flame
3	800	16	50:1	>95	Flame
6	800	16	50:1	>95	Flame
10	400	20	20:1	>95	Flame

[a]Measurement: Perkin-Elmer, Model 420, with either graphite furnace HGA500 accessory, or an air–acetylene flame.
From Allen *et al.*[8] with permission.

Table 24 Inter-laboratory comparative test, DIN, October 1978.

	Element				
	Cd	Co	Cu	Ni	Pb
Number of laboratories	10	10	10	10	10
Outliers	1	–	–	1	–
Arithmetic mean of concentrations found (μg litre^{-1})	2.8	16	126	20	28
Standard deviation					
Absolute (μg litre^{-1})	0.19	1.6	10	1.7	3.5
Relative (variance) (per cent)	6.7	9.9	7.9	8.6	12.4

From Allen *et al.*[8] with permission.

In Table 23 are presented results obtained by this procedure in the determination of cadmium. In all cases recoveries are better than 95 per cent.

An inter-laboratory comparative test organised by DIN (German Institute for Standardisation) involved the determination of five metals in potable water, spiked with five heavy metals; ten laboratories participated in this test. The results are given in Table 24. Application of the Dixon test for 5 per cent significance level led to the rejection as outliers of one cadmium value and one nickel value. After elimination of these outliers the variance for lead was 12.4 per cent and for the other metals below 10 per cent.

The sodium diethyldithiocarbamate extraction method has been one of the most widely used techniques of preconcentration for trace metal analysis by atomic absorption spectrometry.[10–20] This extraction method can be generally classified into two major categories. The first one involves the extraction of metal dithiocarbamate complexes into oxygenated organic solvents such as methyl isobutyl ketone and then analysing the solvents directly.[10–14] The other one is to extract the metal complexes into oxygenated or chlorinated organic solvents such as chloroform, methyl isobutyl ketone, etc., followed by a nitric acid back-extraction, and then analysing the trace elements in an acid solution. The latter category has been the subject of a number of reports.[15–19] There are several drawbacks associated with the acid back extraction of metal dithiocarbamates. The kinetics are generally slow and the efficiency of acid extraction is poor for certain metals such as cobalt, copper and iron.[17]

The Analytical Quality Control Committee of the Water Research Centre, UK[20] has organized a comparative study, involving 11 participating laboratories in the UK, of the accuracy of determining cadmium, copper, lead, nickel, and zinc in river waters. Two laboratories used a preconcentration technique involving the use of sodium diethyldithiocarbamate dissolved in chloroform, eight laboratories used a concentration by evaporation, and one laboratory used direct aspiration without preconcentration. Comparison of precisions obtained by chelation solvent extraction and evaporative methods (Table 25) suggests that in no case was the precision of the former method inferior. Also, spiking recoveries tend to be higher by the chelation method. In general, with the possible exception of zinc, bias obtained by the solvent extractions is not unexpectedly lower than that obtained by evaporation methods where negative biases predominate presumably due to mechanical losses of metal salts during the evaporation process (Table 26).

Bradshaw *et al.*[21] developed a procedure for the determination of 1–100 mg $litre^{-1}$ copper in natural waters involving complexation with 1-pyrrolidinedithiocarbamate, extraction into methyl isobutyl ketone, and determination by atom-trapping atomic absorption spectrometry.

Regan and Warren[22] preconcentrated lead by extraction of the aqueous sample with a 1 per cent solution of ammonium tetramethylenedithiocarbamate in 4-methylpent-2-one. The organic phase is then used for the direct determination of lead by graphite furnace atomic absorption spectrometry.

Chakraborti *et al.*[23] determined traces of cadmium, cobalt, copper, iron, nickel, and lead at the concentrations found in natural waters by extracting with a carbon tetrachloride solution of sodium diethyldithiocarbamate. The extract was evaporated to dryness and the residue mineralized in one drop of concentrated nitric acid prior to analysis by graphite furnace atomic absorption spectrometry.

Comparative data for the dithiocarbamate preconcentration AAS analyses of metals from various types of non-saline waters is presented in Table 27.

Inductively coupled plasma atomic emission spectrometry (ICPAES) Tao *et al.*[36] preconcentrated 100-fold traces of cadmium, cobalt, chromium, iron, manganese, molybdenum, nickel, lead, vanadium, and zinc in river and sea-water by extraction with ammonium pyrrolidinedithiocarbamate and hexamethyleneammonium hexamethylenedithiocarbamate dissolved in xylene, followed by inductively coupled plasma emission spectrometry of the solvent extract. The detection limits for the elements based on three times the standard deviation of the blank signals and a 100-fold concentration factor range from 0.017 μg $litre^{-1}$ (cadmium) to 0.52 μg $litre^{-1}$ (lead). The detection limits of this extraction method are sufficient for the determination of metals in natural waters although higher sensitivity would be desirable for cadmium, cobalt, lead, and chromium(VI).

The linearity of the calibration graph after 100-fold concentration

Table 25 Precision tests, WRC analytical quality control scheme.

Lab No.	River Water			Spiked river water			
	Conc. found range (μg litre^{-1})	St range (μg litre^{-1})	RSD range (per cent)	Conc. found range (μg litre^{-1})	St range (μg litre^{-1})	RSD range (per cent)	Spike recovery range (per cent)
Cadmium (target St 5 per cent of concentration or 0.25 μg litre^{-1} whichever is greater)							
1,2,3,5,7,8,9 (evaporation)	−0.4–1.0	0.04–0.26	17–87	5.2–3000	0.27–3.10	1.0–5.4	93.1–101.0
4,6 (chloroform–diethyldithiocarbamate)	0.2–0.04	0.04–0.18	20–45	11.8–30.0	0.05–0.33	1.6–2.7	98.0–13.8
Copper (target St 5 per cent of concentration or 2.5 (μg litre^{-1} whichever greater)							
1,2,3,5,7,8,9	2.7–10.6	0.21–1.10	4.3–25.9	17.6–317.3	1.13–7.19	1.3–6.4	90.3–104.3
4,6	2.6–9.8	0.29–0.67	6.8–11.1	12.9–23.4	0.42–0.75	3.2–7.1	98.4–101.7
Lead (target as for copper)							
1,2,3,5,7,8,9	0.2–18.5	0.46–2.27	6.2–92.0	25.6–332.6	0.99–12.8	1.2–7.6	92.8–104.9
4,6	10.2–33.1	1.001.52	4.6–9.8	20.9–87.5	1.52–2.47	2.8–7.3	98.0–106.9
Nickel (target as for copper)							
1,2,3,5,7,8,9	0.4–12.8	0.33–1.00	3.1–115.0	25.5–306.0	0.07–6.66	1.9–3.7	91.6–101.5
4,6	2.5–9.7	0.52–0.64	6.6–20.8	14.4–36.4	0.58–1.13	3.1–4.0	96.4–119.6
Zinc (target as for copper)							
1,2,3,5,7,8,9	3.3–324.5	0.28–4.25	1.3–28.8	19.6–968.1	0.74–11.7	1.2–5.4	91.3–109.2
4,6	14.3–24.1	1.09–1.63	6.8–7.6	25.1–77.4	2.12–2.48	3.2–8.4	95.9–108.4

From Water Research Centre, UK.[20]

Table 26 Bias tests WRC analytical quality control scheme.

Lab. No.	River water 1		River water 2	
	Determined value (μg litre^{-1})	Maximum possible bias (per cent)	Determined value (μg litre^{-1})	Maximum possible bias (per cent)
Cadmium (target 10 per cent of determined or 0.5 μg litre^{-1} whichever greater)				
1,2,3,5,7,8,9 (evaporation)	2.19	−0.13 to 0.73	5.19	−5.13 to 11.82
4,6 (chloroform–diethyldithiocarbamate)		−0.12 to 0.36	5.19	4.7 to 10.29
Copper (target 10 per cent of determined greater) concentration or 5 μg litre^{-1} whichever greater)				
1,2,3,5,7,8,9	24.74	−3.00 to 3.45	54.33	−17.88 to 10.54
4,6		−4.42 to 2.97		−12.63 to 5.07
Lead (target as for copper)				
1,2,3,5,7,8,9	24.22	−5.49 to 3.10	51.28	−19.74 to 13.06
4,6		−1.70 to 5.47		3.72 to 5.00
Nickel (target as for copper)				
1,2,3,5,7,8,9	22.76	−3.25 to 5.48	59.21	−20.18 to 20.63
4,6		−2.89 to 3.70		−6.66 to 23.29
Zinc (target as for copper)				
1,2,3,5,7,8,9	20.15	−3.07 to 4.20	50.31	−12.55 to 9.54
4,6		−4.15 to 6.57	50.31	−14.83 to 9.42

From Water Research Centre, UK.[20]

Table 27 Complexing agent–solvent extraction systems for the dithiocarbamate preconcentration–AAS analysis of metals.

Organic complexing agent	Sample solvent	Type of water sample	Metal	Analytical finish	Detection limit ($\mu g\ litre^{-1}$)	Ref.
Ammonium pyrrolidinedithiocarbamate	Methyl isobutyl ketone	Natural	Pb, Cd	AAS	–	24
Ammonium pyrrolidinedithiocarbamate	Methyl isobutyl ketone	Brackish water	Co, Ni, Cu, Zn, Cd, Pb	AAS	–	25
Ammonium pyrrolidinedithiocarbamate	Methyl isobutyl ketone	Ground water	As^{III}	AAS	0.7	26
Ammonium pyrrolidinedithiocarbamate	Methyl isobutyl ketone	Potable	Cd	AAS	–	27
				AAS	–	28
Ammonium pyrrolidinedithiocarbamate	–	Natural	Cu, Ni, Fe, Co, Cd, Zn, Pb			
Ammonium pyrrolidinedithiocarbamate	Methyl isobutyl ketone	Potable	Pb	AAS	–	24
Ammonium pyrrolidinedithiocarbamate	–	Potable	Ni	AAS	–	30
Ammonium pyrrolidinedithiocarbamate	2,6-Dimethyl- 4-heptanone	Natural effluents	Misc. metals	AAS	–	31
Sodium diethyldithiocarbamate	Methyl isobutyl ketone	Natural	Pb, Cd	AAS	Pb 10 Cd 0.5	32
Sodium diethyldithiocarbamate	Methyl isobutyl ketone	Sea	Cd, Pb	AAS	–	33
Sodium diethyldithiocarbamate	Methyl isobutyl ketone	Natural	Cu, Pb, Cd, Ag, Ni	AAS	–	34
Sodium diethyldithiocarbamate	Isoamyl alcohol	River	Cd, Fe, Zn, Cu, Mn, Pb	AAS	–	35

was observed from the detection limit up to at least 30 ng ml^{-1} (15 μg) for most elements. For iron, manganese, and molybdenum the linearity extended to 100 ng ml^{-1} (50 μg). The ranges proved to be wide enough to encompass the concentrations found in most natural waters. Deviations from linearity at high concentrations probably arise because the metal carbamates cannot be extracted completely into xylene. However, for only a 20-fold concentration the linearity, as expected, extends to about five times the values for 100-fold concentration. The relative standard deviations were 3–5 per cent for all the elements except for cadmium, cobalt, and lead, the lack of precision for which arises from the very low concentrations in natural waters; but even for these elements the recommended method might be applicable to polluted waters.

Other workers[37-39] have employed dibenzylammonium dibenzylthiocarbamate dissolved in 2-ethyl hexyl acetate to preconcentrate cations prior to inductively coupled plasma atomic emission spectrometry, achieving detection limits down to 20 μg $litre^{-1}$.

Spectrophotometry Extraction of chelates by a cholroform solution of sodium diethyldithiocarbamate followed by a spectrophotometric finish has been used to determine total heavy metals[40] and zinc.[41]

Neutron activation analysis Lo *et al.*[42,43] have used chelation with lead diethyldithiocarbamate in chloroform to preconcentrate mercury from samples containing 1–1000 μg $litre^{-1}$ prior to neutron activation analysis. As well as considerably increasing the sensitivity of the analytical procedure this step eliminates interference from sodium and bromine in the water samples. Irradiation was carried out with a neutron flux of 2×10^{12} n cm^{-2} sec^{-1} for 30 h. After cooling for 12 h the 77.6 kev ^{197}Hg gamma peak was assayed with a 38 cm^{-3} Ge(li) detector connected to a 4096 channel pulse height analyser.

Other elements which have been preconcentrated by the solvent extraction of their diethyldithiocarbamates followed by neutron activation analysis include iridium and gallium in amount down to 1 ng $litre^{-1}$ [44,45] and lead.[46]

X-Ray flourescence spectrometry Preconcentration of selenium by ammonium pyrolidinedithiocarbamate followed by energy dispersive X-ray fluorescence spectrometry enabled Marcie[47] to determine down to 10 μg $litre^{-1}$ selenium in natural water.

Coprecipitation with sodium chethyldithiocarbamate has been used to preconcentrate 0.02–0.1 μg levels of iron, manganese, zinc, copper, cadmium, arsenic, lead, and zinc prior to determination by X-ray fluorescence spectrometry.[48]

X-Ray spectrometry A methyl isobutyl ketone solution of ammonium pyrrolidinedithiocarbamate has been used to preconcentrate various metals in amounts down to 250 μg litre^{-1} in natural water prior to determination by X-ray energy spectrometry.[49]

Adeljou and Brown[50] preconcentrated cadmium in natural waters as its dithiocarbamate by extraction with Freon and subsequently back-extracting into an acidic aqueous medium for determination by anodic stripping voltammetry. The use of a preconcentration procedure considerably reduced the overall time required for the determination. Direct determination of cadmium concentrations of 0.1 μg litre^{-1} or more was possible using a calibration graph; for lower concentrations the use of the standard addition method was necessary. The minimal amount of cadmium that could be determined reliably was 0.025 μg litre^{-1}.

Gas chromatography Rigin and Yurtaev[51] determined bis(trifluoroethyl)dithiocarbamate complexes of heavy metals in natural water samples, after extractive gas chromatographic separation, by atomic fluorescence. To filtered samples (100 ml) were added ammonium bis(trifluoroethyl)dithiocarbamate and ammonia solution. The mixture was refluxed at 60 °C for 15 min before cooling, acidifying (pH 2–5), and extracting the complexes with 2 ml carbon tetrachloride. The organic extract was dried and the heavy metal complexes separated by capillary gas chromatography with helium as carrier gas. The gas leaving the column was injected into the atomizer of a multi-element atomic fluorescence spectrometer. Metal concentrations were obtained from calibration curves. River water samples containing organic complexing agents (humic acids or surfactants) were pretreated with ozone. Detection limits were slightly inferior to those obtained using atomic absorption but were in good agreement with those obtained from NBS standard water samples.

2.1.2 *Dithizone*

Atomic absorption spectrometry Ihnat *et al.*[52] determined copper, zinc, cadmium, and lead in natural fresh waters by a preconcentration method based on preconcentration into an *n*-butyl acetate solution of dithizone, 8-hydroxyquinoline, and acetylacetone and compared the results obtained with those found by four other methods based on direct electrothermal atomization atomic absorption spectrometry, heat evaporation followed by flame atomic absorption spectrophotometry, and differential pulse anodic stripping voltammetry.

In the chelation method 100 ml aliquot of the sample was preconditioned with 20 ml 5 per cent ammonium tartrate and adjusted in the presence of *p*-nitrophenol indicator to pH 6 with (1 + 1) ammonium hydroxide or tartaric acid crystals. The sample was saturated with 2 ml *n*-butyl acetate then extracted for 1 min with 8 ml of extractant, which contained 0.4 g dithizone, 6.0 g 8-hydroxyquirioline, and 200 ml acetyla-

Table 28 Detection limits of analytical methods.

	Detection limit[a] (μg litre^{-1})			
	Cu	Zn	Cd	Pb
Heat evaporation/flame atomic absorption spectrometry	0.8	0.3	0.1–0.4	1.6
Electrothermal atomization/atomic absorption spectrometry with Perkin-Elmer heated graphite atomizer	0.5	–	0.01	0.2
Differential pulse anodic stripping voltammetry with hanging mercury electrode and mercury film electrode	0.1 0.05	– –	0.05 0.001	0.05 0.005
Electrothermal atomization/atomic absorption spectrometry with Varian Techtron carbon rod atomizer	0.1	0.05	0.005	0.05
Solvent extraction/flame atomic	0.5–1.6	0.3–1.1	0.5–1.1	2.1–5.1

[a]Detection limits usually defined as 2 × or 3 × standard deviation of replicate analyses of reagent blanks and low level samples; ranges reflect differences in technique (Evap/FAAS) or different runs (Solv. ext/FAAS).
From Ihnat *et al.*[52]

cetone in 1 litre *n*-butyl acetate. The aqueous phase was discarded after separation for 20 min and the organic phase was brought to 10 ml with extractant. Organic solutions were aspirated into an air/acetylene flame of an atomic absorption spectrometer and absorbances were read after 4 s integration. Non-atomic absorption was corrected simultaneously by a hollow cathode hydrogen lamp. Concentrations of elements in samples were calculated from linear regression equations of absorbances against concentrations of standards.

Detection limits of the five analytical methods are presented in Table 28. Other applications of dithizone to the preconcentrations of cations are summarized in Table 29.

2.1.3 Diantipyrylmethane

Emission spectography Petrov *et al.*[58] assessed the suitability of 0.05 M solution of diantipyrylmethane in chloroform or dichloromethane as an extractant for preconcentrating traces of 20 metals in mine waters to which ammonium thiocyanate had been added. This reagent does not complex with nickel, aluminium, iron or manganese but can be used for the preconcentration of copper, zinc, vanadium, tin, molybdenum,

Table 29 Preconcentration of cations by extraction with solvent solutions of dithizone.

Solvent	Type of water	Elements	Finish technique	Detection limit	Reference
Ethyl propionate	Natural	Ag, Be, Cd, Co^{II}, Ni, Zn, Fe^{II}, Pb^{II}, Al	AAS	–	53
Carbon tetrachloride	Snow	Ag	GFAAS	0.5 ng $litre^{-1}$	55
Methyl isobutyl ketone	Natural	Au, Ag	AAS	Ag 5 pmol Au 2 pmol	54
Chloroform	Natural	Hg	AAS	–	56
Methyl isobutyl ketone	Natural	Hg	AAS	–	57

niobium, bismuth, tungsten, gadolinium, cobalt, cadmium, and antimony in the presence of 0.2–5.0 mg $litre^{-1}$ of iron.

For determination of the microcomponents, 200–500 ml of mine water, acidified with nitric acid and filtered through a membrane filter with pore diameter of 0.45 μm, containing not over 1 mg of total iron, was boiled for 20 min with 0.5 g of ammonium persulphate. Then the sample was evaporated to a volume of 30–40 ml, cooled and quantitatively transferred to a separatory funnel. Then 1.4 ml of concentrated sulphuric acid, 5 ml of 6.5 M ammonium thiocyanate solution, and 0.5 g of ascorbic acid were added for reduction of the iron and copper, 20 ml of 0.05 M diantipyrylmethane solution in chloroform was added and the trace elements were extracted for 15 min. The extraction was repeated and the extracts were combined and placed in a crucible. The solvent was carefully removed under a lamp, the sample was incinerated with 2 ml of sulphuric acid on a plate to remove sulphur trioxide and then in a muffle furnace at 500–600 °C for 20 min. After the calcination the residue was mixed with 80 mg of spectrographic base (composition C:K_2SO_4:Ni = 2:1:0.002) and transferred to the depression of a carbon electrode. The analytical lines (nm) of the elements were: Co 341.2, Zn 328.2, Cu 327.4, V 318.5, Sn 317.5, Mo 317.0, Nb 316.3, Bi 306.7, Ga 287.4, W 294.7, Cd 228.8, Sb 231.5, the reference element was Ni, 324.3, 313.4, 298.1, 231.1 nm.

2.1.4 Thenoyltrifluoroacetone

Atomic absorption spectrometry Methyl isobutyl ketone solutions of thenoyltrifluoroacetone have been used[59] to preconcentrate, by factors of up to 20, traces of manganese as the Mn^{II} thenoyltrifluoroacetone complex in natural water samples prior to atomic absorption spectrometry.

Apparatus:

Perkin Elmer 403 atomic absorption spectrophotometer equipped with a standard air–acetylene burner head or equivalent. The instrumental settings used are summarized in Table 30.

Reagents:

Standard manganese solutions: 1.00 mg ml^{-1} stock, prepare by dissolving 0.100 g of pure manganese metal in 10 ml of 7 M nitric acid and dilute to exactly 100 ml with water.

Buffer solution (pH 9.5). Add ammonia solution (1 M) to 1 M ammonium chloride to adjust the pH to 9.5 (pH meter).

Hydroxylamine hydrochloride solution, 2.5 per cent w/v.

Thenoyltrifluoroacetone solution, 0.01 M, dissolve 0.88 g in 400 ml of methyl isobutyl ketone.

Procedures:

Iron was removed by methyl isobutyl ketone extraction before the thenoyltrifluoroacetone extraction. About 94 per cent of the manganese remained in the aqueous phase after a single methyl isobutyl ketone

Table 30 Atomic-absorption conditions for manganese determination.

Wavelength	279.5 nm
Lamp current	20 mA
Slit width	1 mm (0.7 nm spectral band width)
Burner height	at position 14
Burner slot	0.5 × 104 mm
Air:	
pressure	30 psig
flow rate	24.0 litre min^{-1}
Acetylene:	
pressure	8 psig
flow rate	6.2 litre min^{-1}
Sample feed rate	3.0 ml min^{-1}
Read out time	11 s
Lamp	Intensitron manganese hollow cathode lamp

From Kato.[59]

extraction. 10 mg of iron can be separated satisfactorily from 0.1–5.0 μg of manganese, by this procedure.

Removal of iron by methyl isobutyl ketone extraction: Before the thenoyltrifluoroacetone extraction, adjust the sample volume to 40 ml and add 40 ml of concentrated hydrochloric acid. Mix well, and then add 10 ml of methyl isobutyl ketone and shake vigorously for 1 min. Drain off and transfer the aqueous phase to a 200 ml quartz dish. To decompose residual organic matter, add 2 ml of concentrated nitric acid and heat gently for a while. Add 1 ml of 60–70 per cent perchloric acid and evaporate the solution until white fumes appear (i.e. nearly to dryness). Dilute the residue to *c.* 40 ml with water, they carry out the thenoyltrifluoroacetone extraction.

Thenoyltrifluoroacetone extraction: Take not more than 40 ml of the sample solution containing 0.08–200 μg of manganese in a separatory funnel. Add 1 ml of hydroxylamine hydrochloride solution and mix well. Add 5 ml of buffer solution and dilute to 50 ml with water, mix well. Then, using a Thymol Blue pH test paper, adjust the pH to 9.5 with dilute ammonia solution if necessary. Add 5.0 ml of thenoyltrifluoroacetone solution and shake vigorously for 1 min. Let the solution stand for 20 min then drain off and discard the aqueous phase (containing a small amount of white bubbles). Aspirate the organic phase into an oxidizing airacetylene flame and measure the atomic absorption (decribed below as absorbance, for convenience) at 279.5 nm. Construct a calibration curve by taking 0 (corresponding to reagent blank), 0.10, 0.25, 0.50, 1.25, and 2.50 μg of manganese, then proceed as described above for the thenoyltrifluoroacetone extraction. Run a reagent blank through the entire procedure.

Chromium, iron, hafnium, niobium, nickel, rhodium, tin, titanium and zirconium interfere strongly in this method even at low concentrations and most other metals (except strontium) give low results when resent at high concentrations, but 10 mg each of silver, arsenic(V), barium, beryllium, cadmium, caesium, germanium, mercury(II), iridium, lanthanum, molybdenum(VI), lead, palladium(IV), rubidium, selenium(IV), tellurium(VI), thallium(I), and vanadium(V), and 20 mg each of potassium and sodium did not interfere. Anionic interferences were restricted to cyanide, fluoride, and thiocyanate; 10 mg each of bromide, carbonate, perchlorate, iodide, nitrate, sulphate, and thiosulphate did not interfere, nor did borate (54 mg), chloride (20 mg), citrate (950 mg), phosphate (30 mg), and tartrate (730 mg). However, addition of tartrate to the aqueous phase before the extraction did not prevent the interferences.

Scintillation counting Strontium-89 and strontium-90 have been preconcentrated by extraction at pH 10.5 with 2-thenoyltrifluoroacetone trioctylphenyl oxide in cyclohexane.[60] Strontium activity in the extract was determined by liquid scintillation counting. Testemale and Leredde[61] preconcentrated strontium-90 using 2-thenoyltrifluoroacetone and tributylphosphate dissolved in carbon tetrachloride. The organic extract was back-extracted with a small volume of aqueous nitric acid prior to counting.

2.1.5 Volatile metal chelates suitable for gas chromatography

Various workers[62–68] have preconcentrated selenium(IV) from natural waters (500 ml) by converting it to its 4,6-dibromopiazselanol or 1,2-diamino-3,5-dibromopiazselanol and extracting this with 1 ml toluene prior to gas chromatography. Down to 0.002 μg litre^{-1} selenium(IV) and total selenium can be determined by this procedure.

Cobalt has been determined in amounts down to 4×10^{-11} by the method described below involving chelation with 6,6,7,7,8,8,8-heptafluoro-2,2-dimethyl-3,5-octanedione (H fod) followed by gas chromatography using an electron capture detector (Ross *et al.*[69]).

Transfer an aliquot (100 ml) of the aqueous sample containing approximately 1×10^{-4} g cobalt to a Pyrex tube fitted with a screw cap. Place a small Tefloncoated stirring bar in the culture tube which is positioned over a stirrer hot plate. Turn on the stirring mechanism before the addition of the reagents, add 0.1 ml aliquot of 0.5 N sodium hydroxide, followed by 1.0 ml of 0.1 M, 1,3-H fod dione (H fod) in benzene (present in excess to ensure complete reaction of cobalt), and 0.2 ml of 35 per cent hydrogen peroxide (to ensure cobalt is completely coverted to the trivalent state). Place a small piece of Teflon tape over the mouth of the tube and screw a cap on tightly to ensure no leakage of vapour during the reaction. Place the reagents in water bath heated to 84–86 °C and allow to react for 30 min with continuous stirring. The temperature should not be increased much in excess of 75 °C to avoid the formation of H fod dihydrate which does not readily gas chromatograph. Allow the reagents in the tube to separate into two layers and decant the organic layer and put into a screw top vial. Add a 1 ml aliquot of 0.1 N sodium hydroxide to the benzene layer for back-washing. A thick, white precipitate forms immediately. This precipitate is the sodium salt of H fod, which is not sufficiently soluble to be entirely dissolved in this volume of aqueous layer. Add 4 ml of distilled water and shake the contents vigorously. The precipitate usually disappears, however, if some remains, washing with distilled water will remove it. The benzene layer can now be decanted and is now ready for gas chromatography.

The gas chromatograph instrument conditions for analysis are listed below:

Instrument, Hewlett Packard Model 402.

Column, 2 ft × ¼ in o.d. Pyrex tube packed with 5 per cent Dow Corning LSX-3-0295 on 60–80 mesh Gas Chrom P.

Column temperature, 135 °C.

Helium eluant, 60 ml min^{-1}.

Auxiliary gas, 95 per cent argon and 5 per cent methane, 100 ml min^{-1}

Injection port temperature, 135 °C.

Detector temperature, 180 °C.

Known aliquots of the benzene solutions are injected into the chromatograph. Peak heights are obtained and compared with the calibration curve produced from the analyses of a standard cobalt 6,6,7,7,8,8,8-heptafluoro-2,2,-dimethyl-3,5-octanedionate solutions. When these particular conditions are used, the *cis* and *trans* isomers of cobalt 6,6,7,7,8,8,8-heptafluoro-2,2-dimethyl-3,5-octanedionate are both eluted as one peak, thereby simplifying the analysis.

Beryllium has been preconcentrated[70] at the μg $litre^{-1}$ level by extraction of 200 ml of sample with 1 ml of a benzene solution of 1,1,1-trifluoro-2,4-pentadione followed by electron capture gas chromatography of the extract.

The copper(II) and nickel(II) complexes of bis(acetylpivalylmethane) ethylenedi-imine (H_2[en $(APM)_2$] show sufficient volatility and thermal stability to allow for their successful gas chromatography. Although somewhat less volatile than complexes such as 1,1,1-trifluoroacetylacetonate, the chelates of H_2^-[en$(APM)_2$] possess a remarkable thermal stability and can be eluted undecomposed from a gas chromatograph at temperatures as high as 300 °C. The relative ease with which the copper(II) and nickel chelates of H_2[en$(APM)_2$] were formed in aqueous solution and their facile extraction into a wide range of water-immiscible organic solvents indicated to Belcher *et al.*[71] the possible usefulness of this ligand in quantitative analysis for copper and nickel. Only with this ligand was the separation of the copper(II) and nickel chelates by gas chromatography consistently reproducible and effective enough for quantitative purposes.

To preconcentrate copper and nickel take the aqueous sample (100 ml) (buffered to pH 7.0) containing copper and nickel (0.02–0.12 mg Ml^{-1}) in vials, and add 1 M sodium acetate solution (1 ml) to each to give pH 8.0. A solution of H_2[en$(APM)_2$] in *n*-hexane (1 ml) was added and the mixture was heated on a boiling water bath for 15 min, add cyclohexane (2 ml), i.e. 50-fold preconcentration, to the cooled mixture and seal the vials. Extract the copper and nickel chelates by shaking the vials for 1 h on the mechanical shaker. Subject portions (1–5 μl) of the organic phase to flame ionization gas chromatography with the column compris-

Table 31 Application of miscellaneous chelating agents to the preconcentration of cations in natural waters.

Chelating agent	Extraction solvent	Type of water	Elements	Finish technique	Detection limit ($\mu g\ litre^{-1}$)	Ref.
1-(2-pyridylazo)naphthol	Benzene and isobutyl methyl ketone	Natural	Zn	AAS	–	72
1-(2-pyridylazo)naphthol	Benzene	Natural	Zn	Spectrometric	–	73, 74
1-(2-pyridylazo)naphthol	Chloroform	Natural	Cr^{III}	Spectrometric	5.0	
8-hydroxyquinoline	Chloroform	Natural	Cr^{III}	Spectrometric	–	75
2-mercaptobenzobenzthiazole	Butyl acetate	Natural	Zu	AAS	0.02	76
Heptoxime	Methanol/ toluene	Natural	Ni	Differential pulse polarography	1	77
Benzoin-oxine	Chloroform	Natural	Mo	Spectrophotometric	0.1	78
Benzoin-oxine	Methyl isobutyl ketone	Natural	W	AAS	–	79
Tri-iso-octyl phosphorothioate	Methyl isobutyl ketone	Stream water	Ag	AAS	0.0002	80
Monoiso-octylmethyl phosphonate	None	Natural	Yttrium	Spectrometry	–	81
Octyl α anilobenzyl phosphonate	Chloroform	Natural	Zu, Cu	Spectrometry	–	82

Tributyl phosphate	Toluene	Natural	^{233}U	Liquid scintillation counting	–	83
Tributyl phosphate	Sodium hydroxide back–extraction	Potable	^{94}Tc	Gas flow proportional counting	–	84
N-M-tolyl-z-methyloxybenzo-hydroxamic acid	Chloroform	Natural	V^{V}	Spectrophotometric	–	85
N-phenyl-2-naphthohydroxamic acid	Chloroform	Natural	U^{VI}	Spectrophotometric	–	86
Nicotinichydroxamic acid–trioctylmethyl ammonium cation	Methyl isobutyl ketone	Natural and waste	Mn	AAS	2	87
6-methyl-3-methyl-2-[4-N-methyl-anilophenylazo] benzthiazolium chloride	Benzene/tributyl phosphate	Natural	Zn	Spectrophotometric	–	88
Nerolic (5-amino-2-anilino-benzene sulphonic acid)	Toluene	Potable	V^{V}	Spectrophotometric	–	89
Sulpharazen (5-nitro-2-3-(4*p*-sulphophenylazophenyl)-1-trizeno) benzenearsonic acids	Toluene/amyl alcohol	Natural	Zn, Pb	Spectrophotometric	Zn 5×10^{-3} Pb 0.1	90
1,3-diaminothiourea (3-thiocarbohydrazide)	Benzene and chloroform, isopropyl ether and ethyl acetate, and amyl acetate and amyl alcohol	Natural	Hg	–	–	91

Table 31 cont.

Chelating agent	Extraction solvent	Type of water	Elements	Finish technique	Detection limit (μg litre^{-1})	Ref.
Mesityl oxide	–	Natural	V^{V}	Spectrophotometric	–	92
5,7-dichloro-8-hydroxyquinoline	Butyl acetate	Natural	V	AAS	10	93
5-nitro-*o*-phenylene diamine	Toluene	Natural	Se		–	94
Phenylacetic acid	Chloroform	Natural	U^{VI}	Titration	–	95
			Cd	Aniodic stripping voltammetry	–	96
Hexahydroze pinium hexahydroazepine-1-carbothioate	Misc.	Natural	Mn^{II}, Fe^{II}, Co, Ni, Zn, Pb^{II}, Cu^{II}	AAS	Mn 6 Fe 40 Co 30 Ni 20 Zn 10 Pb 100 Cu 20	97
Benzylamine/pelargonic acid	Water/decane	Natural	Cu, Zn, Fe, Cd, Pb	AAS	Fe 3 Cu 1 Zn 2 Pb 5	98
Tropalone	Toluene	Natural	Sn	AAS	–	99
Sb^{III} Iodide complex with *N*, *N*-diphenylbenzamidine	Chloroform	Industrial waste	Sb^{III}	Spectrophotometrically	200	100
Trilaurylamine-*N*-oxide	None	Natural	Hg	AAS	–	101

Ammonium thiocyanate	4(-5-noxyl) pyridine	Natural	Cu	AAS	2	102
Napthoquinone thiosemicarbazone	Methyl isobutyl ketone	Natural	Cu	AAS	–	103
Dithiocarbamate-methyllithium (conversion to $PbMe_4$)	Chloroform	Natural	Pb	GFAAS	5	104
Iodide complex Acetyl	Toluene	Natural	^{75}Se	AAS	< 100	105
acetonate Trifluoroacetyl	Chloroform	Natural	Be	AAS	–	106
acetonate	Cyclohexane	Natural	Be	GLC	0.02	107
4-(4-diethylaminophenyl) 930-2,5 dichlorobenzene sulphonate and benzo-18-crown-6 reagents	Benzene:2,5-dichlorobenzene (1:1)	River	K	Spectrophotometric by flow injection analysis	–	256
Mixed chelates	Hexane	Natural	Cu, Fe, Co, Cd, Pb, Zn	AAS	–	108
Misc. chelates	Misc. solvents	Natural	Mo	–	–	109
Misc. chelates	Misc. solvents	Sea	Mo	AAS	–	110–114
Misc.	–	Sub-surface waters	Misc. metals	Spectrophotometric, AAS, Neutron activation analysis	–	115
None	Methyl isobutyl ketone	Nuclear waste processing solutions	K	Spectrophotometric using 2-(5-bromo-2-pyridylazo)-5-diethylamino)phenol by flow injection analysis	257	

ing 5 per cent silicone gum rubber E-350 or Universal B support maintained at 260 °C and a carrier gas flow rate of 10 ml min^{-1}.

2.1.6 *Other chelating agents*

A limited amount of work has been carried out on the application of other types of chelating agents in the preconcentrations of cations. This is summarized in Table 31.

2.1.7 *Comparison of chelating agents*

Chambers and McClellan[116] evaluated several organic complexing agents and solvents for the extraction of copper, cadmium, antimony, arsenic, and selenium from water samples prior to their determination by atomic absorption flame emission spectrometry (Table 32). The object of this exercise was to select the best system for each element. Two millilitres of organic extractant per 200 ml water were used throughout this study, i.e. preconcentration factor of 100.

It was necessary to optimize the atomic absorption instrumental variables for each metal solvent pair as these variables will influence the analytical sensitivity. The variables considered were fuel flow rate,

Table 32 Complexing agents and concentrations.

Metal	Complexing agent	Concentration
Copper	Cupferron	1.0 per cent
	Diethyldithiocarbamates	0.01 M
	1(-2-pyridylazo)napthol	0.1 per cent
Cadmium	Dithizone	0.01 per cent
	1(-2-pyridylazo)napthol	0.1 per cent
	8-Hydroxyquinoline (oxine)	0.1 M
	8-Isopropyltropolone	0.1 per cent
	1-Nitroso-2-napthol	0.1 per cent
Antimony	Cupferron	1.0 per cent
	Diethyldithiocarbamate	1.0 per cent
	Ammonium pyrrolidinedithiocarbamate	0.2 per cent
	Tri-*n*-octylamine	5.0 per cent
Arsenic	Tri-*n*-octylamine	5.0 per cent
	Trioctyl phosphine oxide	0.1 M
	Ammonium pyrrolidinedithiocarbamates	0.2 per cent
	Diethyldithiocarbamate	1.0 per cent
Selenium	Diethyldithiocarbamate	1.0 per cent
	Ammonium pyrrolidinedithiocarbamate	0.2 per cent

From Chambers and McClellan.[116]

oxidant gas flow rate, burner height, and lamp current. In general, the optimum oxidant flow rate did not vary a great deal from solvent to solvent and the optimum range was fairly broad. The combustion characteristics of the solvent and the type of flame required for atomization of the element determine the optimum fuel flow rate. The optimum range in most cases was narrow. The optimum lamp current for each metal does not vary with the solvent. Therefore, it is necessary to optimize this variable only once for each metal. Table 33 lists the optimum instrumental settings for each metal studied.

Several organic solvents were evaluated for enhancement effects on copper, cadmium, antimony, and selenium. Table 34 lists the solvents used and the enhancement values obtained for each of the metal ions. Those solvents giving the greatest sensitivity enhancement, which are insoluble in water, were then used in solvent extraction studies.

Good enhancement of copper sensitivity is obtained using esters, ketones, aromatic hydrocarbons, aldehydes, and others. Isopropyl acetate, *n*-butyl acetate, 2-heptanone, toluene, butyraldehyde, and *n*-butyl ether were considered the most suitable for copper extraction studies. Alcohols depress the absorbance readings for copper because of their high viscosity. Acetate esters, ketones, and aldehydes were found

Table 33 Optimum instrumental conditions.

Metal	Solvent	Acetylene flow rate (litre min^{-1})	Air flow rate (litre min^{-1})	Burner height (mm)	Lamp current (mA)
Copper	Isopropyl acetate	1.61	6.94	3.0	8.0
	Butyl acetate	2.32	6.39	1.0	8.0
	Butyraldehyde	1.94	6.39	4.0	8.0
	2-Hepotanone	1.94	5.23	1.0	8.0
	n-Butyl ether	2.68	5.82	1.5	8.0
	Toluene	11.61	6.39	1.0	8.0
Cadmium	Methyl isobutyl ketone	1.94	5.23	1.5	4.0
	n-butyl acetate	1.61	5.82	1.0	4.0
	Cyclohexanone	2.32	6.94	0.0	4.0
	Butyraldehyde	1.94	3.49	2.0	4.0
Antimony	Methyl isobutyl ketone	1.94	6.39	2.5	15.0
	n-butyl acetate	2.32	6.39	1.0	15.0
	2-Octanone	3.04	6.94	1.0	15.0
		Hydrogen flow rate	Nitrogen flow rate		
Arsenic	Water	30.0	10.8	5.5	18.0
Selenium	Water	31.2	10.8	4.0	12.0

From Chambers and McClellan.[116]

Table 34 Enhancement values with various solvents.

Solvent	Enhancement (A_0/A_{sq})			
	Cu	Cd	Sb	Se
n-Butyl acetate	1.37	1.32	1.58	0.21
Isopropyl acetate	2.00	1.44	–	0.29
Butyl butyrate	1.58	0.77	1.33	–
Ethyl acetoacetate	1.51	1.05	1.13	–
Methyl benzoate	1.09	0.44	0.68	–
Butyraldehyde	2.09	1.77	0.62	–
2,4-Pentanedione	1.35	1.16	1.52	–
2-Heptanone	1.77	1.35	1.05	0.26
Methyl isobutyl ketone	1.70	1.35	1.75	0.30
Methyl ethyl ketone	1.80	1.27	1.53	–
Cyclohexanone	1.29	1.05	1.36	–
2-Octanone	1.58	1.07	1.50	–
n-Butanol	1.25	0.81	1.18	–
n-Hexanol	0.45	0.65	0.87	–
n-Octyl alcohol	0.58	–	–	–
Cyclohexanol	0.29	–	–	–
p-Xylene	1.80	–	–	–
Toluene	1.83	–	–	–
Nitrobenzene	0.83	–	–	–
p-Dioxane	1.38	–	–	–
Propylene carbonate	1.12	0.72	1.01	–
n-Butyl ether	2.32	–	–	–
Amyl acetate	–	–	1.22	–
n-Pentanol	–	–	1.00	–

From Chambers and McClellan.[116]

to give the best enhancement of cadmium sensitivity. Alcohols depress cadmium sensitivity. Butyraldehyde, *n*-butyl acetate, methyl isobutyl ketone, and cyclohexanone were chosen for cadmium extraction studies. Enhancements for antimony are highest with acetate esters and ketones. Methyl isobutyl ketone, *n*-butyl acetate, and 2-octanone give high enhancements of antimony sensitivity and are insoluble in water. These solvents were used in the antimony extraction studies. A suitable flame could not be obtained with aromatic hydrocarbons and *n*-butyl ether solutions of antimony.

The atomic absorption determination of arsenic and selenium presents a problem due to their low resonance lines. The most sensitive line for selenium is at 1960.3 Å while the most sensitive line for arsenic is at 1937.0 Å. These wavelengths are in the low ultraviolet region where many gases, including oxygen, absorb over wide bands. Consequently, studies carried out in this region should be made with no air in the light path. Since nitrogen does not absorb at wavelengths above 1850 Å, measurements may be more accurately made in the vacuum ultraviolet region and by sweeping all the air from the flame region with nitrogen. Acetylene and coal-gas fuels are impractical for arsenic and selenium determinations because the flame species absorb strongly in the low ultraviolet region. Absorbance readings for aqueous arsenic solutions were found to be 2.43 times larger in a nitrogen shielded air entrained hydrogen flame than in an air–acetylene flame. Selenium sensitivity is enhanced by a factor of 2.54 by using a nitrogen shielded hydrogen flame.

Organic solvents give poor results with both arsenic and selenium. The absorbance readings are lower than for aqueous solutions and the noise level is considerably larger. Therefore, arsenic and selenium determination are best made by extracting into an organic solvent such as carbon tetrachloride, or chloroform, followed by back-extraction into an aqueous solution for preconcentration purposes.

The best extraction system for atomic absorption determination of copper is the *n*-butyl ether–1-(2-pyridylazo)naphthol system (Table 35). Quantitative extraction is obtained over a wide pH range and the

Table 35 Sensitivity and extraction efficiency for copper.

Solvent	Chelating or complexing agent	Maximum absorbance	Extraction (per cent)	Suggested pH range
Butyl acetate	Cupferron	0.478	100	3.50–8.60
	Diethyldithiocarbamate	0.465	100	2.50–5.00
	1-(-2-pyridylazo)napthol	0.450	98	3.20–5.50
n-Butyl ether	Cupferron	0.515	100	2.65–9.60
	Diethyldithiocarbamate	0.555	98	1.00–7.00
	1-(-2-pyridylazo)napthol	0.640	100	6.00–9.90
Butyraldehyde	Cupferron	0.500	98	3.80–5.20
	Diethyldithiocarbamate	0.442	90	1.00–4.10
	1-(-2-pyridylazo)napthol	0.500	98	3.00–4.20
Toluene	Cupferron	0.480	100	4.30–9.55
	Diethyldithiocarbamate	0.520	98	6.50–7.50
	1-(-2-pyridylazo)napthol	0.510	100	4.70–9.20
2-Heptanone	Cupferron	0.500	100	2.60–5.00
	Diethyldithiocarbamate	0.435	80	1.00–2.50
	1-(-2-pyridylazo)napthol	0.520	97	3.20–9.00
Isopropyl acetate	Cupferron	0.620	100	2.70–4.50
	Diethyldithiocarbamate	0.620	98	3.20–5.07
	1-(-2-pyridylazo)napthol	0.630	93	5.55–6.57

From Chambers and McClellan.[116]

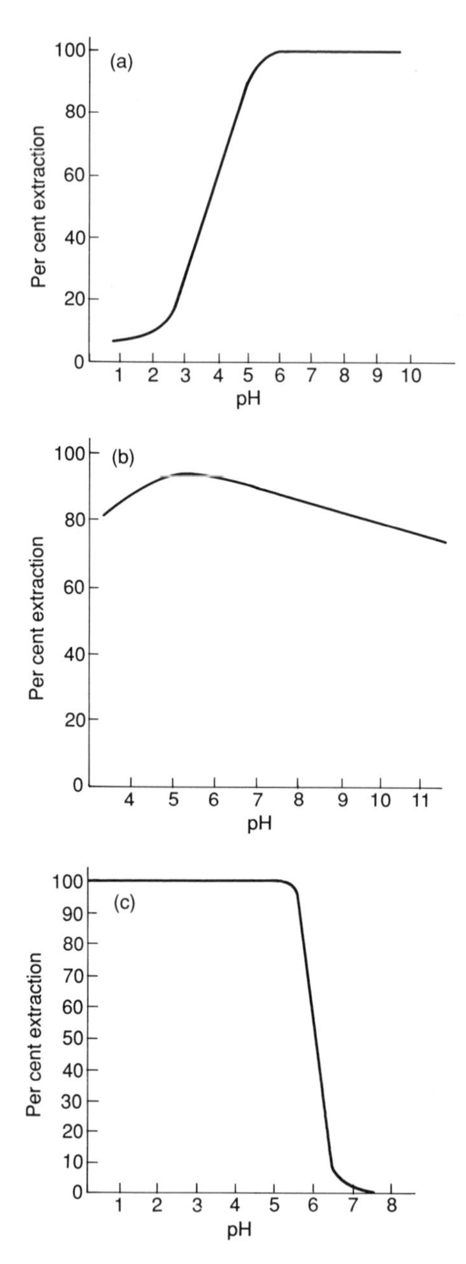

Fig. 11 Per cent extraction versus pH curve: (a) for copper using an *n*-butyl ether–PAN system, (b) for cadmium using an MIBK–dithizone system, (c) for selenium using an APDC–chloroform system. From Chambers and McClellan[116] with permission.

sensitivity is exceptionally high. Fig. 11(a) shows the per cent extraction versus pH curve for this system.

In cadmium determinations severe clogging of the aspiration system occurred with the *n*-butyl acetate–oxine extraction system. If the oxine concentration was lowered to the point that a good aspiration rate is maintained, less than 10 per cent extraction of cadmium is obtained. The methyl isobutyl ketone–oxine extraction was carried out using a 1 per cent oxine solution and only 8 per cent extraction resulted. Highest absorbance readings were obtained with a methyl isobutyl ketone–dithizone extraction system. A per cent extraction/pH curve for this extraction system is shown in Fig. 11(b).

Ammonium pyrrolidinedithiocarbamate can be used to extract antimony from hydrochloric acid. The extraction is quantitative with a hydrochloric acid concentration of 0.001–1.00 M. Methyl isobutyl ketone or *n*-butyl acetate may be used as solvents but methyl isobutyl ketone is more sensitive for atomic absorption determinations.

The atomic absorption sensitivity of arsenic and selenium is best in aqueous solutions. Enhancement of sensitivity is obtained by concentrating the elements by solvent extraction, followed by back-extraction into an aqueous system. Extractions of arsenic with diethyldithiocarbamate or ammonium pyrrolidinedithiocarbamate followed by back-extraction with copper gives quantitative recovery of the arsenic. Selenium is also extracted by ammonium pyrrolidinedithiocarbamate in chloroform. Quantitative extraction is obtained over a pH range of 1.5–5.30. Figure 11(c) shows a per cent extraction versus pH curve for this system. Quantitative back-extraction is achieved using a 50 mg $litre^{-1}$ cyanide ion solution.

Table 36 Detectability limits by solvent extraction–atomic absorption based on aqueous phase concentration.

Metal	Extraction system	Absorbance	Metal ($\mu g\ litre^{-1}$)
Copper	1-(-2-pyridylazo)napthol-*n*-butyl ether	0.010	0.01
Cadmium	Dithizone–methyl isobutyl ketone	0.008	0.10
Antimony	Ammonium pyrrolidinedithiocarbamate–methyl isobutyl ketone	0.020	10
Arsenic[a]	Diethyldithiocarbamate–chloroform	0.011	20
	Ammonium pyrrolidinedithiocarbamate–chloroform	0.010	20
Selenium[a]	Ammonium pyrrolidinedithiocarbamate–chloroform	0.005	2

[a] Back extraction was performed and measurement was made in aqueous medium.
From Chambers and McClellan.[116]

The above solvent complexing agent systems were used to prepare a standard curve for each element in the low μg litre^{-1} range in order to determine linearity and detection limit. Copper, cadmium, and antimony standard curves were prepared by extracting 200 ml of aqueous solution with 2 ml of organic solvent and aspirating the organic phase. Arsenic and selenium standard curves were prepared by extracting 200 ml of aqueous solution with 10 ml of organic solvent. The organic phase was then back-extracted with 2 ml of the appropriate aqueous solution and aqueous solution aspirated. In all cases, the aqueous and organic phases were presaturated with each other. Table 36 lists the detectability limit of each metal by this method. The detectability limits is defined as the lowest concentration which will give an absorbance reading twice the noise level. Good linearity and reproducibility was obtained in all cases.

2.2 Metal cations – sea-water

Generally speaking the sensitivity requirements for the analysis of sea-water are greater than those for natural waters. Consequently, in addition to spectrophotometric and atomic absorption methods, a wide variety of other sensitive techniques used in conjunction with preconcentration have been used for the determination of cations in sea-water. These include inductively coupled plasma atomic emission spectrometry, X-ray fluorescence spectrometry, neutron activation analysis, anodic stripping voltammetry, and other techniques.

2.2.1 Dithiocarbamic acid derivatives

Spectrophotometric method Yang *et al.*[133] have described a spectrophotometric method for the determination of dissolved titanium in sea-water after preconcentration using sodium diethyldithiocarbamate.

Atomic absorption spectrometry The dithiocarbamate extraction method has been one of the most widely used techniques of preconcentration for trace metal analysis by atomic absorption spectrometry.[137–146] This extraction method can be generally classified into two major categories. The first one comprises conversion of the metals to metal–dithiocarbamate chelates, then the extraction of the metal–dithiocarbamate complexes from a large volume of the aqueous phase into a smaller volume of oxygenated organic solvents such as methyl isobutyl ketone (thereby achieving concentration of metals), and then analysing the solvents directly.[137–141] The other one is to extract the metal complexes into oxygenated or chlorinated organic solvents such as chloroform, methyl isobutyl ketone, etc. followed by a nitric acid back-extraction, and then analysing the trace elements in the acid solution. The latter category has been the subject of a number of reports.[142-146] There are several drawbacks associated with the acid back-extraction of metal dithiocarbamates, the kinetics is generally slow and the efficiency of acid extraction is poor for certain metals such as cobalt, copper, and iron.[144] Dithiocarbamate systems can simultaneously extract manganese as well as other trace metals under suitable conditions.[147-149]

Statham[150] has optimized a procedure based on chelation with ammonium pyrrolidinedithiocarbamate and diethylammonium diethyldithiocarbamate for the preconcentration and separation of dissolved manganese from sea-water prior to determination by graphite furnace atomic absorption spectrometry. Freon-TF was chosen as solvent because it appears to be much less toxic than other commonly used chlorinated solvents, it is virtually odourless, has a very low solubility in sea-water, gives a rapid and complete phase separation and is readily purified. The concentrations of analyte in the back-extracts are determined by graphite furnace atomic absorption spectrometry. This procedure concentrates the trace metals in the sea-water 67.3-fold. When a 350 ml sea-water sample was spiked with ^{54}Mn and taken through the chelation, extraction, and back-extraction procedures, the observed recovery of the radio-tracer was 100.6 per cent.

Burton[151] has also described an atomic absorption method for the determination of down to 0.3 nmol litre^{-1} manganese in sea-water. Samples for the analysis of manganese were pressure filtered through 0.4 μm nucleopore filters. To 350 ml filtrate, 20 ml of an aqueous solution of the complexing agents (2 per cent w/v in both ammonium and diethylammonium diethyldithiocarbamate) were added, and the solution extracted first with 35 ml and then with 20 ml Freon for 6 min. The combined extracts were shaken with 100 μl of concentrated nitric acid for 30 s. After standing for 5 min, 5 ml distilled water was added and the solution shaken for 30 s. The aqueous phase was separated and combined with that from a further back-extraction using the same procedure. The combined aqueous solutions were returned to the shore laboratory and manganese determined by electrothermal atomic absorption spectrophotometry.

Cadmium, copper, and silver have been determined by an ammonium pyrrolidinedithocarbamate chelation followed by a methyl isobutyl ketone extraction of the metal chelate from the aqueous phase[152-153] followed by graphite furnace atomic absorption spectrometry. The detection limits of this technique for 1 per cent absorption were determined to be: Cu, 0.03 μmol litre^{-1}; Cd, 2 nmol litre^{-1}; Ag, 2 nmol litre^{-1}.

Moore[154] used the solvent extraction procedure of Danielson *et al.*[155]

to determine iron in frozen sea-water. To a 200 ml aliquot of sample was added 1 ml of a solution containing sodium diethyldithiocarbamate (1 per cent w/v) and ammonium pyrrolidinedithiocarbamate (1 per cent w/v) in 1 per cent ammonia solution and 0.65 ml 1 M hydrochloric acid to bring the pH to 4. The solution was extracted three times with 5 ml volumes ov 1,1,2-trichloro-1,2,2-trifluorethane and the organic phase evaporated dryness in a silica vial and treated with 0.1 ml Ultrex hydrogen peroxide (30 per cent) to initiate the decomposition of organic matter present. After an hour or more, 0.5 ml 0.1 M hydrochloric acid was added and the solution irradiated with a 1000 W Hanovia medium pressure mercury vapour discharge tube at a distance of 4 cm for 18 min. The iron in the concentrate was then compared with standards in 0.1 M hydrochloric acid using a Perkin-Elmer Model 403 Spectrophotometer fitted with a Perkin-Elmer graphite furnace (HGA 2200).

The coefficient of variation of analyses was 21 per cent for seven subsamples containing 1.6 nmol Fe $litre^{-1}$ and 30 per cent for eight subsamples at 0.6 nmol Fe $litre^{-1}$. The detection limit was estimated to be 0.2 nmol Fe $litre^{-1}$. The efficiency of the extraction procedure was tested using sea-water spiked with iron-59, which indicated a recovery of 97 per cent and with stable iron of 86 per cent.

Apte and Gunn[156] have described a method for the preconcentration and determination of copper, nickel, lead, and cadmium in small samples of estuarine and coastal waters by liquid–liquid extraction and electrothermal atomic absorption spectrometry. The metals, in 1.25 ml samples, were chelated with ammonium pyrrolidinedithiocarbamate and extracted into 1,1,1-trichloroethane; all the chemical stages were carried out in sample cups of a graphite furnace atomic absorption spectrometer. Detection limits for copper, cadmium, lead, and nickel were 0.3, 0.02, 0.7, and 0.5 μg $litre^{-1}$, respectively.

Filippelli[157] determined mercury at the subnanogram level in seawater using graphite furnace atomic absorption spectrometry. Mercury(II) was concentrated using the ammonium tetramethylenedithiocarbamate (ammonium pyrrolidinedithiocarbamate, APDC)–chloroform system, and the chloroform extract was introduced into the graphite tube. A linear calibration graph was obtained for 5–1500 ng of mercury in 2.5 ml chloroform extract. Because of the high stability of the Hg^{II}–APDC complexes, the extract may be evaporated to obtain a crystalline powder to be dissolved with a few microlitres of chloroform.

About 84 per cent of mercury was recovered in a single extract (97 per cent in two extractions). The calibration graph was prepared by plotting the peak height against amount of mercury added to 500 ml distilled water. The optimized experimental' conditions are as follows: lamp current, 6 mA; wavelength, 253.63 nm; drying 100 °C for 10 s; ashing, 200 °C for 10 s; atomization, 2000 °C for 3 s; and purge gas, nitrogen 'stopped flow'. The coefficient of variation of this method was about 2.6 per cent at the 1 μg litre^{-1} mercury level. The calibration graph is linear over the range 5–1500 μg mercury.

Lo *et al.*[158] have developed a new method of back-extracting metals from their solution as a metal chelate in an organic solvent. This procedure uses dilute mercury(II) solution instead of nitric acid. This back-extraction method is based on the fact that the extraction constant of the mercury(II) ammonium pyrolidinedithiocarbamate complex is much greater than most of the common trace metals of environmental importance. The substitution of mercury(II) for other metals in the form of dithiocarbamate complex is extremely fast and the efficiency of recovery is nearly 100 per cent for a number of metals including cobalt, copper, and iron. In addition, the back-extracted solution contains a low concentration of mercury(II) which is virtually interference free in graphite furnace atomic absorption spectrometry due to its high volatility. This two-step preconcentration method preconcentrates a number of trace metals such as cadmium, cobalt, copper, iron, manganese, nickel, lead, and zinc in sea-water by graphite furnace.

ICPAES Sugimae[159] developed a method for lead, zinc, cadmium, nickel, manganese, iron, vanadium, and copper in which they chelated with diethyldithiocarbamic acid and the chelates extracted with chloroform and the chelate decomposed prior to determination by inductively coupled plasma atomic absorption spectrometry. When 1-litre water samples are used, the lowest determinable concentrations are: Mn, 0.063 g litre^{-1}; Zn, 0.13 μg litre^{-1}; Cd, 0.25 μg litre^{-1}; Fe, 0.25 μg litre^{-1}; V, 0.38 μg litre^{-1}; Ni, 0.5 μg litre^{-1}; Cu, 0.5 μg litre^{-1}; Pb, 2.5 μg litre^{-1}. Above these levels, the relative standard deviations are better than 12 per cent for the complete procedure.

Muyazaki *et al.*[160] found that di-isobutyl ketone is an excellent solvent for the extraction of the 2,4-pyrrolidinedithiocarbamate chelates of cadmium, lead, zinc, iron, copper, nickel, molybdenum, and vanadium from sea-water. Unlike halogenated solvents, it does not produce noxious substances in the inductively coupled plasma, has a very low aqueous solubility and gives 100-fold concentration in one step. Detection limits are 0.02 μg litre^{-1} (cadmium) to 0.60 μg litre^{-1} (lead). The results indicate that the proposed procedure should be useful for the precise determination of metals in oceanic water, although a higher sensitivity would be necessary for lead and cadmium. The relative standard deviations were 4 per cent for all elements except cadmium and lead, which had relative standard deviations of about 20 per cent owing to the low concentrations determined.

Bloekaert *et al.*[161] applied ICPAES with ammonium pyrrolidinethiocarbamate preconcentration to the determination of cadmium, copper, iron, manganese, and zinc in highly saline waste waters. The application of ICPAES to the analysis of brines containing up to 37.5 mmol litre^{-1} sodium chloride has been discussed.[162, 163] Detection limits as low as 4 μg litre^{-1} have been claimed.

Cathodic and anodic scanning voltammetry Van der Berg[164] determined zinc complexing capacity in sea-water by cathodic stripping voltammetry of zinc–ammonium pyrrolidinedithiocarbamate complex by cathodic ions. The successful application of cathodic stripping voltammetry, preceded by adsorptive collection of complexes with ammonium pyrrolidinedithiocarbamate for the determination of zinc complexing capability in sea-water is described. The reduction peak of zinc was depressed as a result of ligand competition by natural organic material in the sample. Sufficient time was allowed for equilibrium to occur between the natural organic matter and added ammonium pyrrolidinedithiocarbamate. Investigations of electrochemically reversible and irreversible complexes in sea-water of several salinities are detailed, together with experimental measurements of ligand concentrations and conditional stability constants for complexing ligands. Results obtained were comparable with those obtained by other equilibrium techniques but the above method had a greater sensitivity.

Brugman *et al.*[165] compared results obtained by anodic stripping voltammetry and atomic absorption spectrometry in the determination of cadmium, copper, lead, nickel, and zinc in sea-water. Three methods were compared. Two consisted of atomic absorption spectrometry but with preconcentration using either Freon or methyl isobutyl ketone and anodic stripping voltammetry was used for cadmium, copper, and lead only. Inexplicable discrepancies were found in almost all cases. The exceptions were the cadmium results by the two methods and the lead results from the Freon with atomic absorption sepctrometric methods and the anodic scanning voltammetric methods.

Clem and Hodgson[214] discuss the temporal release of traces of cadmium and lead in bay water from EDTA, ammonium pyrrolidinediethyldithiocarbamate, humic acid, and tannic acid after treatment of the sample with ozone. Anodic scanning voltammetry was used to determine these elements.

α-Activity Shannon and Orden[166] determined polonium-210 and lead-210 in sea-water. These two elements are extracted from sea-water (at pH 2) with a solution of ammonium pyrrolidinedithiocarbamate in isobutyl methyl ketone (20 ml organic phase to 1.5 litres of sample). The two elements are back-extracted into hydrochloric acid and plated out of solution by the technique of Flynn,[167] but with use of a PTFE holder in place of the Perspex one, and the α-activity deposited is measured. The solution from the plating-out process is stored for 2–4 months, then the plating-out and counting are repeated to measure the build-up of polonium-210 from lead-210 decay and hence to estimate the original ^{210}Pb activity.

X-ray spectrometry Tseng *et al.*[168] determined cobalt-60 in sea-water by successive extractions with tris (pyrrolidinedithiocarbamate) bismuth(III) and ammonium pyrrolidinedithiocarbamate and back-extraction with bismuth(II). Filtered sea-water adjusted to pH 1.0–1.5 was extracted with chloroform and 0.01 M tris(pyrrolidinedithiocarbamate) bismuth(III) to remove certain metallic contaminants. The aqueous residue was adjusted to pH 4.5 and re-extracted with chloroform and 2 per cent ammonium pyrrolidinethiocarbamate, to remove cobalt. Back-extraction with bismuth(III) solution removed further trace elements. The organic phase was dried under infra-red and counted in a germanium/lithium detector coupled to a 4096 channel pulse height analyser. Indicated recovery was 96 per cent, and the analysis time excluding counting was 50 min per sample.

High performance liquid chromatography (HPLC) Boyle *et al.*[255] preconcentrated cobalt from 100 ml samples of surface sea-water using an ammonium pyrrolidinedithiocarbamate–carbon tetrachloride extraction system. Cobalt was determined in the extract by high performance liquid chromatography using a luminal post column Chemiluminescence detection system. The detection limit of this method was 5 pmol cobalt litre^{-1}.

X-Ray fluorescence spectrometry Murata *et al.*[258] give details of equipment and a procedure for determination of traces of heavy metals by solvent extraction using di-isobutyl ketone and isobutyl methyl ketone, combined with microdroplet analysis by X-ray fluorescence spectrometry using a specially designed filter paper, sodium diethyldithiocarbamate is used as chelating agent. The limits of detection for manganese, iron, cobalt, nickel, copper, zinc, and lead were 15, 16, 8, 8, 13, 13, and 40 μg litre^{-1} respectively for a 100 μl sample volume. Table 37 shows that the results are in fair agreement with the reference values determined by atomic absorption spectrometry.

2.2.2 Dithizone

Atomic absorption spectrometry Hirao *et al.*[179] concentrated lead in sea-water using a chloroform solution of dithizone and determined it in amounts down to 40 μg litre^{-1} by graphite furnace atomic absorption spectrometry. Lead in 1 kg acidified sea-water was equilibrated with ^{212}Pb of a known radioactivity, extracted with dithizone in chloroform, back-extracted with 0.1 M hydrochloric acid, and subjected to graphite

Table 37 Results of analyses (μg litre^{-1}) of liquid samples by X-ray fluorescence spectrometry, with reference values obtained by atomic absorption spectrometry (AAS).

Ion analysed	Waste-water						Sea-water	
	Sample A		Sample B		Sample C			
	XRF[a]	AAS	XRF[b]	AAS	XRF[b]	AAS	XRF[b]	AAS
Mn	120	130	–	–	240	240		
Fe	170	200	130	150	100	90	60	60
Co	220	240	–	–	–	–	–	–
Ni	130	140	20	24	70	80	–	–
Cu	–	–	40	40	30	30	20	20
Zn	–	–	140	140	60	50	–	–
Pb	–	–	70	70	50	40	–	–

Sample: concentrated 10-fold times.
[a] DDTC-IBMK extraction.
[b] DDTC-DIBK extraction.
From Murata *et al.*[258]

furnace atomic absorption spectrometry by a two-channel spectrometer. Recovery yield of lead was found to be 60–90 per cent from the radioactivity of ^{212}Pb in the back-extract. Lead concentrations were thus determined with about 10 per cent precision.

2.2.3 8-Hydroxyquinoline

Atomic absorption spectrometry Klinkhammer[169] and Landing[173] have described methods for determining manganese in a sea-water matrix for concentrations ranging from about 30 to 5500 ng litre^{-1}. The samples are extracted with 4 nmol litre^{-1} 8-hydroxyquinoline in chloroform and the manganese in the organic phase is then back-extracted into 3 M nitric acid.[169] The manganese concentrations are determined by graphite furnace atomic absorption spectrophotometry. The blank of the method is about 3.0 ng litre^{-1} and the precision from duplicate analyses is ± 9 per cent (1 so). The theoretical yield of the method is less than 100 per cent since only 80–90 per cent of the aqueous phase is removed after the back-extraction. The actual yield obtained by ^{54}Mn counting was 69.5 ± 7.8 per cent and this can be allowed for in the calculation of results. Environmental Protection Agency standard sea-water samples of known manganese content (4370 ng litre^{-1}) gave good manganese recoveries (4260 ng litre^{-1}).

Atomic absorption spectrometry coupled with solvent extraction of iron complexes has been used to determine down to 0.5 μg Litre^{-1} iron in sea-water.[170,171] Hiire *et al.*[170] extracted iron as its 8-hydroxyquinoline complex. The sample is buffered to pH 3–6 and extracted with a 0.1 per cent methyl isobutyl ketone solution of 8-hydroxyquinoline. The extract is aspirated into an air–acetylene flame and evaluated at 248.3 nm.

Chau and Lum-Shui-Chan[172] investigated the use of atomic absorption in conjunction with solvent extraction using 1 per cent 8-hydroxyquinoline in methyl isobutyl ketone for preconcentration. The detection limit is 3 μg $litre^{-1}$, in which a preconcentration factor of 20 is employed. The disadvantages of the system are that there are interferences, although some of these can eliminated.[213]

X-Ray fluorescence spectrometry Armitage and Zeitlin[174] converted uranium, copper, nickel, cobalt, iron, and manganese to the 8-hydroxyquinolates and extracted these with chloroform. The extract was applied to a filter paper disc in a ring oven at 160 °C and the metals separated prior to final determination by X-ray fluorescence spectrometry. Morris[175] separated microgram amounts of vanadium, chromium, manganese, iron, cobalt, nickel, copper, and zinc from 800 ml sea-water by precipitation with ammonium tetramethylenedithiocarbamate and extraction of the chelates at pH 2.5 with methyl isobutyl ketone. Solvent was removed from the extract and the residue dissolved in 25 per cent nitric acid and the inorganic residue dispersed in powdered cellulose. The mixture was pressed into a pellet for X-ray fluorescence measurements. The detection limit was 0.14 μg or better, when a 10 min counting period is used.

Electron spin resonance spectroscopy Background copper levels in sea-water have been measured by electron spin resonance techniques.[176] The copper was extracted from the sea-water into a solution of 8-hydroxyquinoline in ethyl propionate (3 ml extractant per 100 ml sea-water) and the organic phase (1 ml) was introduced into the electron spin resonance tube for analysis. Signal-to-noise ratio was very good for the four-line spectrum of the sample and of the sample spiked with 4 and 8 ng Cu^{2+}, the graph of signal intensity versus concentration of copper was rectilinear over the range 2–10 μg $litre^{-1}$ of sea-water, and the coefficient of variation was 3 per cent. Traces of copper and lead are separated[177] from macro amounts of calcium, magnesium, sodium, and potassium by adsorption from the sample on to active carbon modified with hydroxyquinoline, dithizone, or diethyldithiocarbamate.

2.2.4 Dimethylglyoxime

Spectrophotometric methods The concentration of nickel in natural waters is so low that one or two enrichment steps are necessary before instrumental analysis. The most common method is graphite furnace atomic absorption after preconcentration by solvent extraction[126] or co-precipitation.[127] Even though this technique has been used successfully for the nickel analyses of sea-water,[128,129] it is vulnerable to contamination as a consequence of the several manipulation steps and of the many reagents used during preconcentration.

This element has been determined spectrophotometrically in sea-water in amounts down to 0.5 μg litre^{-1} as the dimethylglyoxime complex.[130,131] In one procedure[130] dimethylglyoxime is added to a 750 ml sample and the pH adjusted to 9–10. The nickel complex is extracted into chloroform. After extraction into 1 M hydrochloric acid, it is oxidized with aqueous bromine, adjusted to pH 10.4 and dimethylglyoxime reagent added. It is made up to 50 ml and the extinction of the nickel complex measured at 442 nm. There is no serious interference from iron, cobalt, copper, or zinc but manganese may cause low results. In another procedure[131] the sample of sea-water (0.5–3 litres) is filtered through a membrane-filter (pore size 0.7 μm) which is then wet-ashed. The nickel is separated from the resulting solution by extraction as the dimethylglyoxime complex and is then determined by its catalysis of the reaction of tiron and diphenylcarbazone with hydrogen peroxide with spectrophotometric measurement at 413 nm. Cobalt is first separated as the 2-nitroso-1-naphthol complex and is determined by its catalysis of the oxidation of alizarin by hydrogen peroxide at pH 12.4 sensitivities are 0.8 μm litre^{-1} (nickel) and 0.04 μg litre^{-1} (cobalt).

Atomic absorption spectrometry Rampon and Cavalier[180] used atomic absorption spectrometry to determine down to 0.5 μg litre^{-1} nickel in sea-water. Nickel is extracted into chloroform from sea-water (500 ml) at pH 9–10, as its dimethylglyoxime complex. Several extractions and a final washing of the aqueous phase with carbon tetrachloride are required for 100 per cent recovery. The combined organic phases are evaporated to dryness and the residue is dissolved in 5 ml of acid for atomic absorption analysis.

2.2.5 4-(2-Pyridylazo)resorcinol

Spectrophotometric method Nishimura *et al.*[132] described a spectrophotometric method using 2-pyridylazoresorcinol for the determination of down to 0.025 μg litre^{-1} vanadium in sea-water. The vanadium was determined as its complex with 4-(2-pyroiylazo)resorcinol formed in the presence of 1,2-diaminocyclohexane-*N*, *N*, *N′*, *N′*-tetra-acetic acid. The complex was extracted into chloroform by coupling with zephiramine. Difficulties due to turbidity in the chloroform layer and incomplete masking of some cations by 2-pyridylazoresorcinol were overcome by addition of potassium cyanide and washing the chloroform layer with sodium chloride solution. The extinction of the chloroform layer was measured at 560 nm against water as was that of a blank prepared with vanadium-free artificial sea-water. Sixteen foreign ions were investigated and no interferences were found at 5–100 times their usual concentration in sea-water.

Atomic absorption spectrometry Monien and Stangel[178] studied the performance of a number of alternative chelating agents for vanadium and their effect on vanadium analysis by atomic absorption spectrometry with volatilization in a graphite furnace. Two promising compounds were evaluated in detail, namely 4-(2-pyridylazo)resorcinol in conjunction with tetraphenylarsonium chloride and tetramethylenedithiocarbamate. These substances, dissolved in chloroform, were used for extraction of vanadium from sea-water, and after concentrating the organic layer 5 μl were injected into a pyrolytic graphite furnace coated with lanthanum carbide. For both reagents a linear concentration dependence was obtained between 0.5 and 7 μg $litre^{-1}$ after extraction of a 100 ml sample. Using the 2-pyridylazoresorcinol–tetraphenylarsonium chloride system a concentration of 1 μg $litre^{-1}$ could be determined with a relative standard deviation of 7 per cent.

2.2.6 Nitrosophenols

Spectrophotometric methods Various methods have been proposed for the determination of traces of cobalt in sea-water and brines, most necessitating preconcentration. Solvent extraction followed by spectrophotometric measurements[118-125] is the most popular method but has many sources of errors; the big difference in the volumes of the two phases results in mixing difficulties, and the solubility of the organic solvent in the aqueous phase changes the volume of organic phase resulting in decreased reproducibility of the measurements. In many cases, excess of reagent and various metal complexes are co-extracted with cobalt and cause errors in determining the absorbance of the cobalt complex.

The procedure of Kentner and Zeitlin[118] is as follows: to a filtered 750 ml sample of sea-water add 20 per cent aqueous sodium citrate (25 ml), 30 per cent aqueous hydrogen peroxide (1 ml), and 1 per cent ethanolic 1-nitroso-2-naphthol (treated with activated charcoal and filtered) (1 ml), and set aside for 10 min. Extract the cobalt complex into chloroform and back-extract the excess of reagent into basic wash solution (mix 1 M sodium hydroxide (50 ml), 20 per cent aqueous sodium citrate (10 ml), and 30 per cent aqueous hydrogen peroxide (1 ml) with water to produce 100 ml (3 × 5 ml), shaking for 60 s for each extraction. Extract copper and nickel from the organic phase into 2 M hydrochloric acid (5 ml), back-extract any released reagent into basic wash solution (5 ml), and wash the chloroform phase again with 2 M hydrochloric acid (5 ml). Dilute the organic phase to 50 ml (for 30 ml cells) or 30 ml (for

20 ml cells) and measure the extinction at 410 nm against a blank in 10 cm cells.

In another spectrophotometric procedure Motomizu[119] adds to the sample (2 litres) 40 per cent (w/v) sodium citrate dihydrate solution (10 ml) and a 0.2 per cent solution of 2-ethylamino-5-nitrosophenol in 0.01 M hydrochloric acid (20 ml). After 30 min, add 10 per cent aqueous EDTA (10 ml) and 1,2-dichloroethane (20 ml), mechanically shake the mixture for 10 min, separate the organic phase and wash it successively with hydrochloric acid (1: 2) (3 × 5 ml), potassium hydroxide (5 ml), and hydrochloric acid (1: 2) (5 ml); filter and measure the extinction at 462 nm in a 50 mm cell. Determine the reagent blank by adding EDTA solution before the citrate solution. The sample is either set aside for about 1 day before analysis (the organic extract should then be centrifuged) or preferably, it is passed through a 0.45 μm membrane-filter. The optimum pH range for samples is 5.5–7.5. From 0.07 to 0.16 μg litre^{-1} of cobalt was determined; there is no interference from species commonly present in sea-water.

2.2.7 *Pyrocatechol violet*

Spectrophotometric method Korenaga *et al.*[117] have described an extraction procedure for the spectrophotometric determination of trace amounts of aluminium in sea-water with pyrocatechol violet. The extraction of ion-associate between the aluminium/pyrocatechol violet complex and the quaternary ammonium salt, zephiramine (tetradecyldimethylbenzylammonium chloride), is carried out with 100 ml sea-water and 10 ml chloroform. The excess of reagent extracted is removed by back-washing with 0.25 M sodium bromide solution at pH 9.5. The calibration graph at 590 nm obeyed Beer's law over the range 0.13–1.34 μg aluminium. The apparent molar absorptivity in chloroform was 9.8×10^4 litre^{-1} mol^{-1} cm^{-1}.

Several ions—such as manganese, iron(II), iron(III), cobalt, nickel, copper, zinc, cadmium, lead, and uranyl—react with pyrocatechol violet and to some extent are extracted together with aluminium. The interferences from these ions and other metal ions generally present in sea-water could be eliminated by extraction with diethyldithiocarbamate as masking agent. With this agent most of the metal ions except aluminium were extracted into chloroform and other metal ions did not react in the amounts commonly found in sea-water.

The apparent aluminium content of sea-water stored in ordinary containers such as glass and polyethylene bottles decreases gradually, but if the samples are acidified with 0.5 ml litre^{-1} concentrated sulphuric acid the aluminium content remains constant for at least 1 month. Accordingly, samples should be acidified immediately after collection. However, the aluminium could be recovered by acidifying the stored samples and leaving them for at least 5 h.

2.2.8 *Triphenylphosphine oxide*

Spectrophotometric method Korkisch and Koch[134, 135] determined low concentrations of uranium in sea-water by extraction and ion exchange in a solvent system containing trioctylphosphine oxide. Uranium is extracted from the sample solution (adjusted to be 1 M in hydrochloric acid and to contain 0.5 per cent of ascorbic acid) with 0.1 M trioctylphosphine oxide in ethyl ether. The extract is treated with sufficient 2-methoxyethanol and 12 M hydrochloric acid to make the solvent composition 2-methoxyethanol–0.1 M ethereal trioctylphosphine acid–12 M hydrochloric acid (9: 10: 1); this solution is applied to a column of Dowex 1-X8 resin (Cl^{-1} form). Excess of trioctylphosphine oxide is removed by washing the column with the same solvent mixture. Molybdenum is removed by elution with 2-methoxyethanol–30 per cent aqueous hydrogen peroxide–12 M hydrochloric acid (18: 1: 1); the column is washed with 6 M hydrochloric acid and uranium is eluted with molar hydrochloric acid and determined fluorimetrically or spectrophotometrically with ammonium thiocyanate. Large amounts of molybdenum should be removed by a preliminary extraction of the sample solution (made 6 M in hydrochloric acid) with ether.

2.2.9 *Bis(2-ethlhexyl) phosphate*

Spectrophotometric method Flynn[136] has described a solvent extraction procedure for the determination of manganese-54 in sea-water in which the sample with bismuth, cerium, and chromium carriers, is extracted with a heptane solution of bis(2-ethylhexyl) phosphate and the manganese back-extracted with 1 M hydrochloric acid. After oxidation with nitric acid and potassium chlorate, manganese is determined spectrophotometrically as permanganate ion.

2.2.10 *Comparison of chelating agents*

A selection of chelation–solvent extraction methods is summarized in Table 38. It is seen that the majority of these use as the chelating agent, diethyldithiocarbamate, ammonium pyrrolidinedithiocarbamate or a mixture of both. Other chelating agents discussed include dithizone, 8-hydroxyquinoline, and hexahydroazepine-1-carbodithioate. Freon, methyl isobutyl ketone, chloroform, butyl acetate, xylene, and carbon tetrachloride feature as extraction solvents. Detection limits (defined as 2 or 2.5 times the standard deviation of the blank) are in the

Table 38 Preconcentration of metals in-sea water chelation–solvent extraction techniques followed by direct atomic absorption spectrometry and graphite furnace atomic absorption spectrometry.

Metals	Chelating agent	Solvent	Detection limit (μg litre^{-1})		Reference
Direct atomic absorption spectrometry					
Mn, Fe, Co, Ni Zn, Pb, Cu	Hexahydroazepine-1-carbodithioate	Butylacetate	Mn	0.2	182
			Fe	1.5	
			Co	0.6	
			Ni	0.6	
			Zn	0.4	
			Pb	2.6	
			Cu	0.5	
Fe, Pb, Cd, Co, Ni, Cr, Mn, Zn, Cu	Diethyldithiocarbamate	MIBK or xylene			183
Fe, Cu	Ammonium pyrrolidinedithiocarbamate	MIBK	Cu	<1	184
			Fe	<1	
Cd, Zn, Pb, Ca, Ni, Cu, Ag	Dithizone	Chloroform	Ag	0.05	181
			Cd	0.05	
			Zn	0.6	
			Pb	0.04	
			Cu	0.06	
			Ni	0.3	
			Co	0.04	

Cd, Cu, Pb, Ni, Zn	(a)Ammonium dipyrrolidinedithiocarbamate (b)Ammonium dipyrrohidinedithiocarbamate plus diethyldithiocarbamate	MIBK	Cu	10	186
			Cd	2	
			Pb	4	
			Ni	16	
			Zn	30	
Graphite furnace atomic absorption spectrometry					
Cu, Ni, Cd	Ammonium pyrrolidinedithiocarbamate				187
Ag, Cd, Cr, Cu, Fe, Ni, Pb, Zn	Ammonium dipyrrolidinedithiocarbamate	MIBK	Ag	0.02[a]	188
			Cd	0.03	
			Cr	0.05	
			Cu	0.05	
			Fe	0.20	
			Ni	0.10	
			Pb	0.03	
			Zn	0.03	
Cu, Cd, Zn, Ni	Diethyldithiocarbamate plus ammonium pyrrolidinedithiocarbamate	Chloroform	Cu	1.0[b]	189
			Cd	0.2	
			Zn	2	
			Ni	10	
Cd, Pb, Ni, Cu, Zn	Ammonium pyrrolidinedithiocarbamate plus diethyldithiocarbamate	Freon		Not stated	190
Cu	Ammonium pyrrolidinedithiocarbamate	MIBK		<0.5	191
Cd, Cu, Ni, Zn	Dithizone	Chloroform	Cu	0.006[a]	192
			Cd	0.0004	
			Ni	0.032	
			Zn	0.016	

Table 38 cont.

Metals	Chelating agent	Solvent	Detection limit ($\mu g\ litre^{-1}$)		Reference
Cd, Cu, Fe	Ammonium pyrrolidinedithiocarbamate plus diethyldithiocarbamate	Freon	Not stated		193
Cd, Zn, Pb, Cu, Fe Mn, Co, Cr, Ni	Ammonium pyrrolidine-*N*-carbodithioate plus 8-hydroxyquinoline	MIBK	Fe	0.08	194
			Cu	0.10	
			Pb	0.06	
			Cd	0.02	
			Zn	0.34	
Cd	Ammonium pyrrolidinedithiocarbamate	Carbon tetrachloride	Cd	0.006	195 197
Cd, Zn, Pb, Fe Mn, Cu, Ni, Co, Cr	Dithiocarbamate	MIBK	Not stated		196
Cd, Co, Cu, Fe, Mn, Ni, Pb, Zn	Ammonium pyrrolidinedithiocarbamate	Chloroform	Cd	<0.0001[a]	198
			Cu	<0.012	
			Fe	<0.02	
			Mn	<0.004	
			Ni	<0.012	
			Pb	<0.016	
			Zn	<0.08	

Cd, Co, Cu, Fe, Mn, Ni, Pb, Zn	Ammonium pyrrolidinedithiocarbamate	Chloroform	Cd	0.02	199
			Cu	0.24	
			Fe	0.24	
			Mn	0.02	
			Ni	0.08	
			Pb	0.04	
			Zn	1.0	
Mn, Cd	Ammonium pyrrolidinedthiocarbamate and diethylammonium diethyldithiocarbamate	Freon	Mn	0.07[a]	200
		Cd	0.027		
Cd, Cu, Fe, Pb, Ni, Zn	Ammonium pyrrolidinedithiocarbamate and diethylammonium diethyldithiocarbamate	Freon	Not quoted		201

[a] 2 × 5.0 used.
[b] 2.5 × 50 used.

Table 39 Detection limits for metals in sea-water.

	Lowest detection limit reported (μg litre^{-1})	Reference	Highest detection limit reported (μg litre^{-1})	Reference
Manganese	0.004	198	0.2	182
Iron	0.02	198	1.5	182
Cobalt	0.04	185	0.6	182
Nickel	0.012	198	16	186
Lead	0.016	198	4	186
Copper	0.006	192	10	186, 189
Silver	0.02	188	0.05	185
Cadmium	0.0001	198	2	186
Zinc	0.016	192	30	186
Chromium	0.05	188		

ranges shown in Table 39 and as such are often suitable for the analysis of background levels in sea-water. The most sensitive methods for all 10 elements are covered by four references two of which use a chloroform solution of dithizone[185, 192] and two of which use a methyl isobutyl ketone solution or chloroform solution of ammonium pyrrolidinedithiocarbamate.[188, 198]

Sturgeon *et al.*[202] compared five different analytical methods is a study of trace metal contents of coastal sea-water. Analysis for cadmium, zinc, lead, iron, manganese, copper, nickel, cobalt, and chromium was carried out using isotope dilution spark source mass spectrometry (DSSMS), graphite furnace atomic absorption spectrometry (GFAAS), and inductively coupled plasma emission spectrometry (ICPES) following trace metal separation preconcentration (using ion exchange and chelation solvent extraction) and direct analysis (by GFAAS). Table 40 gives results obtained on a sample of sea-water. Overall, there is good agreement in elemental analysis obtained by the various methods.

Although ICPAES is a multi-element technique, its inferior detection limits, relative to graphite furnace atomic absorption spectrometry, would necessitate the processing of large volumes of sea-water, improvements in preconcentration procedures in use up to this time or new alternate preconcentration procedures such as carrier precipitation.

2.2.11 Volatile metal chelates suitable for gas chromatography

Shimoishi[203] determined selenium by gas chromatography with electron capture detection. To 50–100 ml sea-water add 5 ml concentrated

Table 40 Analysis of sea-water samples.

Element	Concentration (μg litre^{-1})			
	GFAAS		ICPES ion exchange	IDSSMS ion exchange
	Direct	Chelation–extraction		
Fe	1.6 ± 0.2[a]	0.5 ± 0.1	1.5 ± 0.6	1.4 ± 0.1
Mn	1.6 ± 0.1	1.4 ± 0.2	1.5 ± 0.1	ND
Cd	0.20 ± 0.1	0.24 ± 0.04	ND	0.28 ± 0.02
Zn	1.7 ± 0.2	1.9 ± 0.2	1.5 ± 0.4	1.6 ± 0.1
Cu	ND	0.6 ± 0.2	0.7 ± 0.2	0.7 ± 0.1
Ni	ND	0.33 ± 0.08	0.4 ± 0.1	0.37 ± 0.02
Pb	ND	0.22 ± 0.04	ND	0.35 ± 0.03
Co	ND	0.018 ± 0.008[b]	ND	0.020 ± 0.003[c]

[a] Precision expressed as standard deviation.
[b] Preconcentrated 100-fold by Chelex 100 ion exchange.
[c] Spark source mass spectrometry, internal standard method.
Weekly retained by Chelex 100 resin not repeated.
From Sturgeon *et al.*[202]

hydrochloric acid and 2 ml 1 per cent 4-nitro-*o*-phenylenediamine and, after 2 h, extract the product formed into 1 ml of toluene. Wash the extract with 2 ml 7.5 M hydrochloric acid, then inject 5 μl into a glass gas–liquid chromatography column (1 × 4mm) packed with 15 per cent of SE-30 on Chromosorb W (60–80 mesh) and operated at 200 °C with nitrogen (53 ml min^{-1}) as carrier gas. There is no interference from other substances present in sea-water.

Measures and Burton[204] used gas chromatography to determine selenite and total selenium in sea-water. Siu and Berman[205] determined selenium(IV) in sea-water by gas chromatography after coprecipitation with hydrous ferric oxide. After coprecipitation, selenium is derivatized to 5-nitropiaz-selenol, extracted into toluene, and quantified by electron capture detection. The detection limit is 5 ng litre^{-1} with 200 ml samples and the precision at the 0.025 μg Se per litre level is 6 per cent.

An example of a gas chromatographic method is that of Lee and Burrell.[206] In this method the aluminium is extracted by shaking a 30 ml sample (previously subjected to ultraviolet radiation to destroy organic matter) with 0.1 M trifluoroacetylacetone in toluene for 1 h. Free reagent is removed from the separated toluene phase by washing it with 0.01 M aqueous ammonia. The toluene phase is injected directly on to a glass column (15 cm × 6 mm) packed with 4.6 per cent of DC 710 and 0.2 per cent of Carbowax 20 M on Gas-Chrom Z. The column is operated at 118 °C with nitrogen as carrier gas (285 ml min^{-1}) and electron-capture detection. Excellent results were obtained on 2 μl of extract containing 6 pg of aluminium.

Isotope dilution gas chromatography–mass spectrometry has also been used for the determination of μg litre^{-1} levels of total chromium in sea-water.[207–209] The samples were reduced to ensure Cr^{III} and then extracted and concentrated as tris(1, 1, 1-trifluoro-2, 4-pentanediono) chromium(III) ($Cr(tfa)_3$) into hexane. The $[Cr(tfa)_2]^+$ The isotopic distribution of mass fragments were monitored into a selected ion monitoring (SIM) mode, (Fig. 12).

Isotope dilution techniques are attractive because they do not require quantitative recovery of the analyte. One must, however, be able to monitor specific isotopes which is possible by using mass spectrometry. In this method, chromium is extracted and preconcentrated from sea-water with trifluoroacetylacetone H(tfa) which complexes with trivalent but not hexavalent chromium. Chromium reacts with trifluoroacetylacetone in a 1:3 ratio to form an octahedral complex $Cr(tfa)_3$. The isotopic abundance of its most abundant mass fragment, $[Cr(tfa)_2]+$ was monitored by a quadrupole mass spectrometer. The isotopic distribution of the $[Cr(tfa)_2]^+$ fragment (*m/e* 358 and 359 here) is evident, in the mars spectrum of $Cr(tfa)_3$. This is readily calculable if the individual elemental abundances are known. Assuming the isotopic abundance of ^{12}C and ^{13}C to be 98.89 per cent, and 1.11 per cent, and ^{50}Cr, ^{52}Cr, ^{53}Cr, and ^{54}Cr to be 4.31, 83.76, 9.55, and 2.38 per cent, respectively, and neglecting any isotopic abundances less than 1 per cent, one can obtain

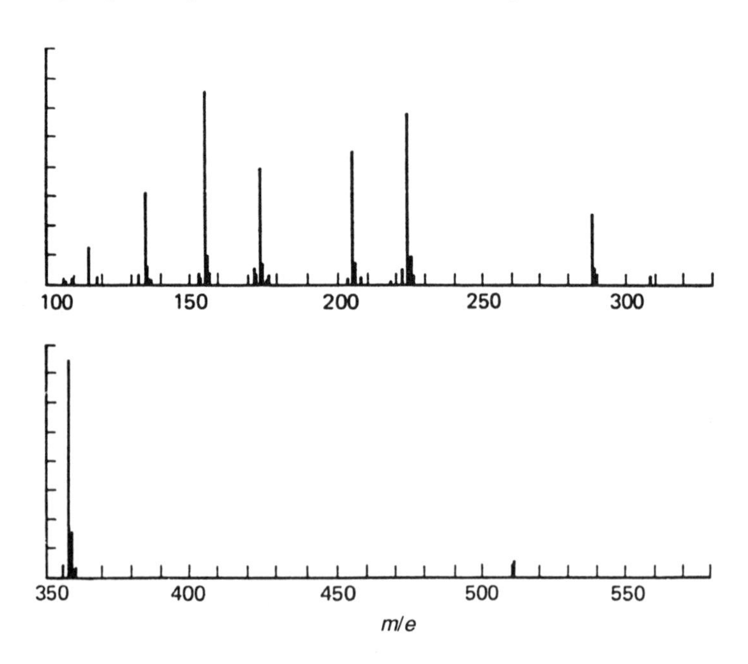

Fig. 12 Mass spectrum of $Cr(ta)_3$. From Lee and Burrell[206] with permission.

a set of calculated abundances for the $[Cr(tfa)_2]^+$ ion. These and the measured isotopic abundances (by SIM) are listed in Table 41. The agreement between the two sets is excellent. The same calculation can be made for the chromium-53 spike solution by using isotopic abundances given by the supplier: ^{52}Cr, 3.44 per cent, ^{53}Cr, 96.4 per cent, and ^{54}Cr, 0.18 per cent. Table 42 lists the calculated and the measured isotopic abundances for the spike solution.

Table 43 shows results of two sea-water sample analyses. Agreement with data obtained by isotope dilution spark source mass spectrometry[210] and graphite furnace[211] was excellent. Lee and Burrell[212] have used a toluene solution of trifluoroacetyl acetone to extract cobalt, iron, indium, and zinc from sea-water.

Table 41 Natural abundance of Cr(tfa).

m/e	Per cent calculated	Per cent measured
356	3.8	3.8
357	0.5	0.6
358	75.2	74.9
359	17.0	16.6
360	3.5	4.0

From Siu *et al.*[207] with permission.

Table 42 Abundance of $[Cr(tfa)_2]^+$ for the chromium-53 spike.

m/e	Per cent calculated	Per cent measured
358	3.1	3.3
359	86.5	86.7
360	9.9	9.1
361	0.5	0.9

From Siu *et al.*[207] with permission.

Table 43 Mean ($\pm$SD) chromium concentration in sea-water (μg litre^{-1}) ($n \geqslant 3$).

ID-GC/MS	ID-SSMS	GFAAS
0.177 ± 0.009	0.17 ± 0.03	0.19 ± 0.03
0.19 ± 0.01[a]	0.18 ± 0.01	ND

[a] Sea-water reference material NASS-1.
ND = not determined.
From Siu *et al.*[207] with permission.

2.3 Anions – non-saline waters

2.3.1 Orthophosphate

Most spectrophotometric methods of phosphate are based on the formation of a heteropoly acid with molybdate. The heteropoly acid formed (molybdophosphate) and its reduction product (so called molybdenum blue) have been used, as has the yellow vanadmolybdophosphoric acid. The protonated forms of these species have also been extracted into organic solvents for spectrophotometry. It well known that molybdophosphate reacts with cationic dyes and organic bases to form ion-pairs. The ion-pair formed can be separated as a precipitate from the aqueous solution.[215-220] Shida and Matsuo[221] reported a very sensitive spectrophotometric method based on formation of the ion-associate of molybdophosphate with methylene blue, flotation of this ion-pair between the aqueous phase and cyclohexane-4-methylpentan-2-one, and displacement of methylene blue from the ion-pair by tetradecyldimethylbenzylammonium ion.

Several preconcentrative extractions of molybdophosphate with cationic dyes have been studied. Safranine T has been used with a mixed acetophenone 1,2-dichlorobenzene solvent.[222,223] Crystal violet or iodine green forms an ion pair extractable with butanol–cyclohexane;[224] the methylene blue ion pair can be extracted with 4-methylpentan-2-one.[225] The ion pair with rhodamine B has been extracted into chloroform–butanol for fluorimetry.[226] Most of these procedures are troublesome because pre-extractions of molybdophosphate or other measures are needed to avoid large reagent blanks.

In an attempt to improve the extractability of ion pairs of molydophosphate and thus the sensitivity of the determination of phosphate, Motomizu *et al.*[227] examined several cationic dyes and extracting solvents. They found that procedures based on ethyl violet and a cyclohexane-4-methylpentan-2-one mixture enabled 10 μg litre^{-1} concentrations of phosphate to be determined spectrophotometrically. In samples of river water and sea-water, the phosphorus content is often in the μg litre^{-1} range. Thus procedure requires small volumes of sample (below 10 ml) and very simple vessels (25 ml test tube) which are easily heated in order to hydrolyse any condensed phosphate. The absorption spectrum of the ion pair formed between molybdophosphate and ethyl violet in the organic phase obtained by the procedure shows a maximum absorbance at 602 nm where the absorbance of the reagent blank is about 0.1. The calibration graphs obtained are linear in the range 0–0.6 μg of phosphorus and the molar absorptivity calculated from the slope of the curve was 2.7×10^5 litre^{-1} mol^{-1} cm^{-1}.

Silicate, vanadate, and tungstate, which may react with molybdate to form heteropoly acids, do not interfere with the determination of phosphorus by the above method even at the amounts of 5×10^{-5} M, 10^{-5} M, and 5×10^{-6} M, respectively. Arsenic(V) causes positive errors because it reacts with molybdate to form the heteropoly acid, which is quantitatively extracted into the organic phase. In natural waters such as river water and sea-water, the arsenic content is very much smaller than the phosphorus content, Tin(II) and (IV) ions at concentrations more than 10^{-6} M interfere; large amounts of tin(II) make it impossible to determine phosphate and tin(IV) causes negative errors.

Phosphorus occurs in natural water as orthophosphate, condensed phosphate (pyro-, and meta-, and poly-phosphate) and organically bound phosphorus, all of which may be present in soluble forms and in suspension. Only orthophosphate can be determined directly by molybdophosphate procedures. Pyro-, tripoly-, and poly-phosphate are completely hydrolysed to orthophosphate by acidification and heating. If samples stored in glass containers are not acidified, the phosphorus content decreases gradually, the decrease becoming significant after 1 day in most cases. River water acidified with 1 ml litre^{-1} of 5 M sulphuric acid usually showed a constant content of phosphorus for about 2 days. Thus if analyses cannot be completed within a few hours of collection, samples should be acidified.

In later work, Motomizu *et al.*[228] point out that in the determination of phosphate at μg litre^{-1} levels in waters the above method has certain disadvantages. First the absorbance of the reagent blank becomes too large for the concentration effect achieved by the solvent extraction to be of much use; for example when 20 ml of sample solution and 5 ml of organic solvent were used, the absorbance of the reagent blank was 0.14. Secondly, the shaking time needed was long and the colour of the extract faded gradually if the shaking lasted more than 30 min. In the course of attempts to improve on this method, they observed that Malachite Green gave a stable dark-yellow species in 1.5 M sulphuric acid (probably a protonated one), whereas Ethyl Violet became colourless within 30 min even in only 0.5 M sulphuric acid.

Malachite green has several advantages over Ethyl Violet.

(1) The absorbance of the reagent blank is very small and 20-fold concentration of phosphate by solvent extraction is possible.

(2) The reagent solution, which consists of Malachite Green, molybdate and 1.5 M sulphuric acid is stable at least for a month.

(3) The method is less troublesome and shaking for 5 min is enough to complete the extraction.

(4) Colour fading in the organic phase does not occur during shaking and standing.

Consequently, the following malachite green preconcentration–spectrophotometric method was developed:

Apparatus:

Hitachi EPS recording spectrophotometer and Hitachi Perrin-Elmer Model 139 spectrophotometer in glass cells of 10 mm path length.

Reagents:

Standard phosphate solution: Dry potassium dihydrogen phosphate at reduced pressure (about 5 mm Hg) and 60 °C to constant weight. Dissolve 0.2772 g of the dried compound in distilled water and dilute to give 1 litre of 2×10^{-3} M solution. For calibration purposes, accurately dilute this solution with distilled water before use.

Extracting solvent: methyl isobutyl ketone.

Reagent solution: dissolve 86 g of ammonium heptamolybdate tetrahydrate in about 900 ml of distilled water add 86 ml of concentrated sulphuric acid, dissolve 0.23 g of Malachite Green (oxalate) and 20 m of tartaric acid in the mixture and dilute to 1 litre with distilled water. About 1 h after mixing, filter the solution through a 0.45 μg membrane filter.

Sulphuric acid, 4.5 M.

Toluene–methyl isobutyl ketone (1: 3) v/v.

Procedure:

Transfer 10 ml of the sample solution containing up to 0.7 μg of phosphorus as orthophosphate, into a 25 ml test tube. Add 1 ml each of 4.5 M sulphuric acid and the reagent solution and mix. Shake the solution with 5 ml of a 1:3 v/v mixture of toluene and methyl isobutyl ketone for 5 min. After phase separation, measure the absorbance of the organic phase at 630 nm against a reagent blank in 1 cm cells.

Most cations and anions commonly found in natural waters do not interfere in this procedure, but arsenic(V) causes large positive errors; arsenic(V) at a concentration of 10 μg litre^{-1} produces an absorbance of 0.070 but can be masked with tartaric acid (added in the reagent solution). When arsenic(V) was present at concentrations of 50 μg litre^{-1}, it was masked with 0.1 ml of 10^{-4} M sodium thiosulphate added after the sulphuric acid. Table 44, shows the results for determination of phosphorus in samples. Recovery tests were done by adding known amounts of phosphate. The results are also shown in Table 44. The recovery of phosphorus was good, 99–103 per cent. The relative standard deviation for phosphorus was 0.6 per cent for 21.0 μg litre^{-1} in sea-water (12 replicates) and 1.1 per cent for 4.3 μg litre^{-1} in potable water (10 replicates).

Table 44 Determination of phosphorus in river and sea-water, and recovery test.

Sample		Phosphorus found ($ng\ ml^{-1}$)	Recovery test				
			Sample taken (ml)	P in sample (ng)	P added (ng)	P found (ng)	Recovery (per cent)
Asahi river	A	6.0	8	48	248	299	101
	B	17.9	8	143	248	387	99
Yoshii river	A	17.5	8	140	248	392	101
	B	16.4	8	131	248	390	103
	C	27.0	8	216	248	459	99
	D	39.4	5	197	248	446	100
Seashore of Seto Inland Sea							
Kojima		46.0	5	230	248	477	100
Shibukawa		17.4	8	139	248	391	101
Tamano		15.3	8	122	248	381	103
Ushimado		12.8	8	102	248	351	100
Nishiwaki		20.9	8	167	248	419	101

From Hotomizu *et al.*[228]

Even more recently Motomizu and Oshima[229] have used the ion-associate formed between molydophosphate and Malachite Green as the basis of a flow injection method for the determination of orthophosphate in river and potable waters in amounts down to 0.1 μg litre^{-1}. The ion-associate was extracted with benzene–4-methylpentane-2-one solvent mixture to achieve the necessary preconcentration. Molybdenum blue solvent extraction procedures have also been applied to polluted waters.[230]

Other molybdenum blue solvent extraction–preconcentration procedures include the use of isoamyl alcohol with a pulse polarographic finish[231] and methyl isobutyl ketone with an inductively coupled plasma atomic emission finish.[232,233] Inductively coupled plasma procedures have the advantage of being applicable to estuarine as well as river waters. Many of the earlier atomic absorption procedures had the disadvantage that arsenics, silicon, and germanium cause positive errors because these elements also form reduced heteropoly acids. Muyazaki *et al.*[232,233] overcome this problem by determining phosphorus at the P^{I} 214.91 nm line as the Mo^{II} 213.61 nm line interfers with the P^{I} 213.62 nm line. To achieve the μg litre^{-1} sensitivity they required they incorporated a 100-fold preconcentration stage into the method.

An inductively coupled plasma atomic emission spectrometer instrument and its operating conditions are described in Table 45. The spectrometer was equipped with 14 fixed slits and a moving slit. It was also capable of selecting different wavelengths by changing the angle of the grating. The wavelength profile of the analytical line was obtained by moving the entrance slit over a small distance.

Reagents:

Use distilled, deionized water throughout.

Phosphorus stock solution (1000 $\mu g\,P\,ml^{-1}$): dissolve potassium dihydrogenphosphate in water. Dilute this solution with water to prepare the working standards just before use.

Molybdate–antimony reagent: dissolve 15 g of ammonium heptomolybdate and 0.189 g of antimony potassium tartrate in water add 180 ml of concentrated sulphuric acid and dilute to exactly 1 litre with water.

Ascorbic acid (reductant) solution (10 per cent).

Ammonia solution.

Phosphate free washing reagent, (Merck Extran MA 03) use for the cleaning of glass ware.

Di-isobutyl ketone, spectroscopic grade.

Extraction procedure:

Transfer 500 ml of sample containing not more than 20 μg of phosphorus to a separatory funnel. Add 60 ml of the molybdate–antimony reagent and mix well. Allow the contents of the funnel to stand for 15 min. Add 2 ml of ascorbic acid solution and leave for 30 min for complete reduction. Extract the reduced molybdoantimonyl phosphoric acid with exactly 5 ml of di-isobutyl ketone by shaking for 5 min. After phase separation, discard the aqueous phase. For samples containing higher concentrations of phosphorus, use 100 ml of sample, 12 ml of the molybdate–antimony reagent, 2 ml of ascorbic acid solution, and 10 ml of di-isobutyl ketone.

Determination of phosphorus by ICPAES:

To avoid plasma instability ignite the plasma and introduce water. Tune the matching network so that the reflected power is a minimum. Increase the reflected power up to about 100 W and introduce di-isobutyl ketone. Tune the matching network again. (The minimum reflected power for diisobutyl ketone was below 5 W.) Allow the plasma system to stabilize for 30 min and measure the P^{I} 214.91 nm emission line intensity while spraying the extract.

The effects of other ions are summarized in Table 46. No interference from silicon and germanium (which cause positive errors in the spectrophotometric method) was observed even at 1000-fold excess. Arsenic(V) was permissible up to 10 times the weight of phosphorus. Above this

Table 45 Instrumental operating conditions.

ICP source	Shimadzu ICPS-2H	
Operating frequency	27 MHz	
Load coil	2-turn copper tubing with teflon coating	
Nebulizer	concentric	
Spectrometer	Shimadzu GEW 170, 1.7 m Ebert	
Grating	2160 line mm^{-1}	
Entrance slit width	30 μm	
Exit slit width	50 μm	
Reciprocal linear dispersion	0.26 nm mm^{-1} (1st order)	
Recording console	Shimadzu RE-7	
Pre-integration time	5 s	
Integration time	20 s	
Attenuation for photomultiplier high voltage	33 divisions	
	DIBK extract	Aqueous solution
Operating power	1.6 kW	1.2 kW
Argon flow rates:		
coolant	13 litre min^{-1}	12 litre min^{-1}
plasma	1.1 litre min^{-1}	0.9 litre min^{-1}
carrier	0.7 litre min^{-1}	0.8 litre min^{-1}
Observation height above load coil	14 mm	16 mm
Sample uptake rate	1.8 ml min^{-1}	1.9 ml min^{-1}

From Motomizu *et al.*[228]

amount, a slight upward background shift was observed which caused a positive error exceeding 5 per cent. However, natural water containing arsenic(V) in amounts more than 10 times that of phosphorus is rare. Most of the other anions did not show any interference and that of nitrite was decreased by addition of sulphamic acid. There was not interference from peroxidisulphate indicating that pretreatment with peroxidisulphate as used for the determination of total phosphorus in water is also applicable to this method.

The calibration graph obtained by the extraction method was linear for at least three orders of magnitude of concentration above the detection limit. The detection limit, defined as the concentration of phosphorus equivalent to three times the standard deviation of the background signal (3α) was 37 μg $litre^{-1}$ phosphorus in the 5 ml of extract. Hence, the detection limit for the original samples (500 ml) was 0.37 μg $litre^{-1}$ phosphorus. Measurements of 5 μg of phosphorus, extracted from 500 ml portions of aqueous solution gave a relative standard deviation of 2.1 per cent.

The values obtained by the inductively coupled plasma method for river water samples are compared with those obtained by the spectro-

Table 46 Effect of other ions on phosphorus results (sample, 100 ml; DIBK, 5 ml, P, 10 μg.

Other ion	Amount added	Recovery (per cent)	Other ion	Amount added	Recovery (per cent)
None	–	100	Fe^{III}	1 mg	101
As^{V}	10 μg	100	Cl^-	2 g	101
	50 μg	102	I^-	10 mg	102
	100 μg	105	Br^-	10 mg	99
As^{III}	1 mg	104	NO_2^-	10 mg	91
Si	10 mg	102		10 mg	104[a]
Ge	10 mg	101	$S_2O_8^-$	1 g	103

[a] 20 mg of sulfamic acid added.
From Motomizu *et al.*[228] with permission.

Table 47 Phosphorus determination in river and sea-water samples (500 ml samples, 5 ml di-isobutyl ketone).

Sample	Phosphorus concentration (ng ml^{-1})			
	ICPAES		Spectrophotometry	
	Range[a]	Average	Range[a]	Average
River water				
A	0.3– 0.4	0.4	0.2– 0.4	0.3
B	5.1– 5.3	5.2	5.0– 5.3	5.2
C	4.3– 4.5	4.4	4.5– 4.8	4.7
D	21.2–22.2	21.9	20.4–22.6	21.5
E	7.4– 8.1	7.7	7.8– 8.5	8.3
F	22.2–23.8	23.1	23.0–26.2	25.0

[a] Range of four measurements
[b] 100 ml of sample; 10 ml of di-isobutyl ketone.
From Motomizu *et al.*[228] with permission.

photometric method in Table 47. They are in good agreement with each other. The precision of the values obtained by ICPAES is better than that achieved spectrophotometrically, especially at low phosphorus concentrations where the absorbance is less than 0.05.

More recently, Muyazaki *et al.*[233] have extended the detection limit of the inductively coupled plasma emission spectrometric technique down to the sub μg litre^{-1} level by measurements of the reduced molybdoantimonyphosphoric acid, not at the phosphorus P^{I} 214.91 nm line, as in the previous method, but at the more sensitive molybdenum(II) 202.03 nm or the antimony(I) 206.83 nm lines. This method is simple, sensitive, and precise. Washing of the organic phase is not necessary because of the low solubility of di-isobutyl ketone in water.

Apparatus and reagents:

The reagents, ICPES instrument, and operating conditions (excepts for the parameters given in Table 48) were identical to those used in the earlier method[232] (Table 45).

Procedure

Extraction: The earlier extraction procedure[232] (see above) was used with the following modification. After the aqueous phase has been discarded wash the stem of the separatory funnel with water and wipe off the droplet remaining with tissue paper. This washing was necessary to prevent contamination of the organic phases with molybdenum or antimony on its transfer from the separatory funnel to a test tube for the ICP measurements.

ICPES measurements: The procedure of tuning the matching network to the di-isobutyl ketone extract is the same as that described above[232] with the exception that no clogging of the nebulizer after spraying water was observed on introduction of the di-isobutyl ketone. The plasma was ignited with an aqueous solution and di-isobutyl ketone was introduced after the matching network had been tuned to water. The Mo^{II} 202.03 nm and Sb^{-I} 206.83 nm emission intensities were measured while the extract was sprayed.

The detection limits of the revised method defined as the concentration of phosphorus equivalent to three times the standard deviations of the background signal, were 5.2 and 45 ng $litre^{-1}$, respectively, for

Table 48 Instrumental operating conditions.

	Molybdenum		Antimony	
	DIBK	Water	DIBK	Water
Operating power (kW)	1.50	1.0	1.40	0.8
Argon flow rate:				
coolant (litre min^{-1})	14.5	13.5	16.0	13.5
plasma (litre min^{-1})	1.25	0.70	1.10	0.75
carrier (litre min^{-1})	0.70	0.75	0.70	0.80
Observation height above load coil (mm)	14	17	14	18
Sample uptake (ml min^{-1})	1.8	1.8	1.8	1.9
Attenuation for photomultiplier high voltage	30	30	31	31

From Motomizu *et al.*[228] with permission.

molybdenum and antimony measurements, when 500-ml water samples and 5 ml of di-isobutyl ketone were used. These values are about 100 and 10 times better than that for the direct phosphorus measurement. The significant improvements in the detection limits arise because the reduced molybdoantimonyphosphoric acid has the composition $PSb_2Mo_{10}O_{40}$ and the Mo II 202.03 nm and Sb I 206.83 nm emission lines are more sensitive than the P I 214.98 nm emission line in ICPES. The relative standard deviations ($n = 10$) of the complete procedure for 1 μg of phosphorus are 2.0 and 2.5 per cent for molybdenum and antimony measurements, respectively. The relative standard deviation of the background signal is 0.3 per cent. The signals for the blank solution are not distinguishable from the background.

Arsenic(V) caused serious interference, but germanium, silicon, and arsenic(II) did not interfere in 100-fold amounts. Nitrite seriously interfers in spectrophotometric methods for molybdenum blue. Molybdenum measurements by the inductively coupled plasma method did not suffer from interference up to 5 mg of nitrite (2500-fold excess over phosphorus). In the antimony measurement, nitrite was permissible up to 0.2 mg (100-fold amount). Above that amount, the interference decreased with increasing amount of antimony in the molybdenum–antimony reagent, although the interference was not completely suppressed. The addition of sulphamic acid decreased the interference from nitrite. The difference in the inteference mentioned above may be related to the mechanism of the reduction of molybdoantimonyphosphoric acid. No interference was observed from peroxidisulphate. Pretreatment with peroxidisulphate as used for the determination of total phosphorus in water, therefore, may be applicable to this method. The results obtained by the molybdenum and antimony measurements method for river

Table 49 Phosphorus determination in river samples.

Sample	Phosphorus concentration (ng ml^{-1})					
	Via Mo		Via Sb		Via P	
	Range[a]	Mean[a]	Range[a]	Mean[a]	Range[a]	Mean[a]
River water						
A	0.49–0.50	0.49	0.4–0.6	0.5	0.4–0.5	0.5
B	0.48–0.50	0.49	0.4–0.5	0.4	0.3–0.4	0.4
C	5.80–6.01[b]	5.91	5.4–6.1	5.8	5.6–5.9	5.7
D	4.55–4.84[b]	4.61	4.0–4.5	4.4	4.3–4.5	4.4
E	4.60–4.94[b]	4.89	4.2–4.8	4.7	5.1–5.3	5.2

[a] Three measurements.
[b] 200-ml sample, 10 ml of DIBK.
From Motomizu *et al.*[228] with permission.

samples are compared with those obtained by direct phosphorus measurements in Table 49. The results are in good agreement.

Bet-Pera *et al.*[234] have described an alternative phosphate preconcentration procedure in which phosphate is converted to 12-molybdophosphoric acid which is then extracted into isobutyl acetate. After evaporation of the organic solvent the complex is dissolved in alkaline solution and after acidification molybdenum(VI) is reduced to molybdenum(III) using a Jones reductor. The resulting molybdenum(III) is reoxidized with iron(III) to molybdate and the resulting iron(II) is determed spectrophotometrically at 562 nm as the iron(II) ferrozine complex. Phosphate in natural water samples was determined by this method in amounts as low as 4 μg litre^{-1} in the final solution with a relative precision of 12 per cent at 2α value.

2.3.2 Silicate

Silicate can be preconcentrated and determined[235] in water in microgram amounts by conversion to molybdosilicic acid in a medium 0.1 N sulphuric acid or hydrochloric acid containing ammonium molybdate. The heteropoly acid is extracted into butanol or isoamyl alcohol and reduced to molybdenum blue using stannous chloride prior to spectrophotometric determination.

2.3.3 Chromate and dichromate

Hexavalent chromium (i.e. chromate and dichromate) is reduced by diethyldithiocarbamate to trivalent chromium,[236, 237] with which it forms an isobutyl methyl ketone soluble complex. Preconcentrated chromate is then determined in the solvent extract by atomic absorption spectrometry at 357.9 nm.

Subramanian[238] studied the factors affecting the determination of trivalent and hexavalent chromium (chromate) by direct complexation with ammonium pyrrolidinedithiocarbamate, extraction of the complex into methyl isobutyl ketone, and determination by graphite furnace atomic absorption spectrometry. Factors studied included the pH of the solution, concentration of reagents, period required for complete extraction, and the solubility of the chelate in the organic phase. Based on the results, procedures were developed for selective determination of trivalent and hexavalent chromium without the need to convert the trivalent to the hexavalent state. For both states the detection limit was 0.2 μg litre^{-1}.

Schaller and Neeb[239] extracted di(trifluoroethyl)dithiocarbamate chelates of hexavalent and trivalent chromium and cobalt from aqueous solution at pH 3 using a carbon-18 column. Adsorbed chelates were eluted with toluene before gas chromatographic analysis with electron capture detection. Broadening of the peaks of cobalt and chromium was reduced by arranging that the end of the capillary was 1 cm within the detector body, the base of which was isolated with glass wool and aluminium foil. The last 0.2 cm of the column was also protected with a sleeve of braided wire. The detection limit for chromium(VI) was 0.05 μg litre^{-1}.

2.3.4 Halides

To preconcentrate fluoride Miyazaki and Brancho[240] converted fluoride into the ternary lanthanum-alizarin complexone fluoride and extracted it into hexanol containing *N,N*-diethylaniline. The extract was analysed directly by inductively coupled plasma atomic emission spectrometry for the determination of fluoride. Measurement of the lanthanum(II) 333.75 nm emission line and comparison with a calibration graph enabled fluoride concentrations as low as 0.59 μg litre^{-1} to be determined in Japanese river water (polluted and unpolluted), coastal seawater, and potable water samples.

Miyazaki and Brancho[241] preconcentrated iodide by conversion to iodine and extraction into xylene. The extract was determined at 172.28 nm using inductively coupled plasma atomic emission spectrometry. Iodate was reduced to iodide which was then treated in the same way to give a total iodide plus iodate content. Detection limits were, respectively, 8.3 and 21 μg litre^{-1} for iodate and iodide. Large concentrations of bromide interfere.

2.3.5 Thiosulphate

Chakraborty and Das[242] have described a method for preconcentrating and determining thiosulphate in photographic waste effluents in which a stable ion-association complex is formed between lead, thiouurea, and thiosulphate in alkaline medium. This complex was extracted into *n*-butyl acetate–butanol (2:1) solvent mixture prior to determination in amounts down to 19 μg litre^{-1} by atomic absorption spectrometry.

2.3.6 Antimonate, trivalent antimony, arsenate, and trivalent arsenic

Metzger and Braun[243] preconcentrated antimonate (pentavalent antimony) by conversion to its chelate with *n*-benzoyl-*n*-phenylhydroxyamine and extraction into chloroform. Trivalent antimony was converted to its chelate with ammonium pyrrolidinedithiocarbamic acid and extraction into methyl isobutyl ketone. Antimony was determined in extracts of rain, snow, and natural water by anodic stripping voltammetry.

To preconcentrate arsenate and antimonate, Mok and Wai[244] and Mok *et al.*[245] first reduced them to their trivalent forms with thiosulphate and iodide at pH 1.0 then adjusted to pH 3.5–5.5, and, in the presence of citrate buffer and EDTA, extracted the trivalent complexes with a chloroform solution of ammonium pyrrolidinedithiocarbamate. Final analysis at the 1 μg litre^{-1} level was achieved by neutron activation analysis.

Stary *et al.*[246] preconcentrated arsenate in natural water by adding tungstate labelled with tungsten-185 and molybdate ions, and extracting the complex formed with a 1:2 dichloroethane solution of tetraphenylarsenium chloride. Down to 0.2 μg litre^{-1} arsenate was then determined in the extract by radiochemical analysis. Dix *et al.*[247] preconcentrated arsenite, plus arsenate, monomethyl arsonate, and dimethyl arsinite by extraction with a cyclohexane solution of thioglycollic acid methyl ester. The four methylthioglycollate derivatives were then determined by temperature programmed capillary column gas chromatography.

Nasu and Kau[248] preconcentrated arsenate, arsenite, and phosphate by a procedure in which a floated (between aqueous and organic phase) ion pair of Malachite Green with molybdophosphate was dissolved by the addition of methanol to the organic layer. Phosphate and arsenate were determined by measuring absorbance of the organic phase. An oxidative (potassium dichromate) or a reductive (sodium thiosulphate) reaction of arsenic was used for the determination of phosphate, arsenate, and arsenite. A positive interference effect was observed in the presence of large amounts of silicon. This was overcome by acidification with concentrated hydrochloric acid. The method was applied to samples of hot spring water, sea-water, and ground water with almost complete recovery of added amounts.

2.4 Anions – sea-water

2.4.1 Iodide and iodate

See Section 2.3.4.

2.4.2 Antimonate, trivalent antimony, arsenate, arsenite, and trivalent arsenic

See Section 2.3.6.

2.5 Organics – sea-water

Aliphatic amines have been determined by a number of methods. Batley *et al.*[249] extracted the amines into chloroform as ion-association complexes with chromate, then determined the chromium in the complex spectrophotometrically with diphenylcarbazide. The chromium might also be determined by atomic absorption spectrometry. With the colorimetric method, the limit of detection of a commercial tertiary amine mixture was 15 μg litre^{-1}. The sensitivity was extended to 0.2 μg litre^{-1} by extracting into organic solvent the complex formed by the amine and Eosin Yellow. The concentration of the complex was measured fluorimetrically.

2.6 Organometallic compounds – non-saline waters

2.6.1 Lead

Chau *et al.*[250,251] preconcentrated low levels of organolead compounds (Me_2Pb^{2+}, Me_3Pb^+, Et_2Pb^{2+}, Et_3Pb^+) in lake water by extraction with a benzene solution of diethyldithiocarbamate, followed by butylation with Grignard reagent to produce the tetrabutyl-lead derivative. Gas chromatography of the extract using an atomic absorption detector enabled down to 0.01 μg litre^{-1} of these substances to be determined in lake water samples.

2.6.2 Mercury

Schintu *et al.*[252] preconcentrated manganese and methylmercury by extraction with dithizone solution. Organic mercury was then recovered from the solvent phase by extraction with aqueous sodium thiosulphate prior to analysis by gold trap cold vapour atomic absorption spectrometry.

2.7 Organometallic compounds – sea-water

2.7.1 Mercury

Graphite furnace atomic absorption spectrophotometry has been used for the determination down to 5 ng litre^{-1} of inorganic and organic mercury in sea-water.[253] The method used a preliminary preconcentration of mercury using the ammonium pyrrolidinedithiocarbamate–chloroform system. Recovery of mercury of 85–86 per cent was reproducibly obtained in the first chloroform extract and consequently it was possible to calibrate the method on this basis. A standard deviation of 2.6 per cent was obtained on a sea-water sample containing 529 ng litre^{-1} mercury. The relative standard deviation of ten repeated determinations of 500 ml distilled water containing 10 ng mercury(II) chloride was 17.4 per cent.

2.7.2 *Tin*

Meinema *et al.*[254] studied the effect of combinations of various solvents with 0.05 per cent tropolone on the preconcentrative recoveries of mono-, di-, and tri-butyltin species either individually or simultaneously present in sea-water. The results obtained by gas chromatography–mass spectrometry after methylation show that Bu_3Sn and Bu_2Sn recoveries appear to be almost quantitative both for neutral and hydrobromic acid-acidified aqueous solutions. BuSn recovery appears to be influenced by the presence of hydrobromic acid in that, in general, recovery rates are higher from solutions acidified with hydrobromic acid than from non-acidified solutions. Bu_3Sn and Bu_2Sn recoveries remain fairly constant with ageing of an aqueous solution of these species over a period of several weeks. BuSn recoveries, however, do decrease with time to a notable extent (20–40 per cent) most likely as a result of adsorption/deposition of BuSn species to the glass wall of the vessel. Addition of hydrobromic acid obviously affects the desorption of these species as recovery of BuSn species increase to almost the same values as obtained from hydrobromic acid-acidified freshly prepared aqueous solutions of BuSn species.

References

1. Kinrade, J. D. and Van Loon, J. C. V. (1974). *Analytical Chemistry*, *46*, 1894.
2. Subramanian, K. S. and Meranger, J. C. (1979). *International Journal of Environmental Analytical Chemistry*, *7*, 25.
3. Bone, K. M. and Hibbert, W. D. (1979). *Analytica Chimica Acta*, *107*, 219.
4. Tessier, A., Campbell, P. G. C., and Bisson, M. (1979). *International Journal of Environmental Analytical Chemistry*, *7*, 41.
5. Smith, J., Nelissen, J., and Van Grieken, R. (1979). *Analytica Chimica Acta*, *111*, 215.
6. Rubio, R. Huguet, J., and Raunret, G. (1984). *Water Research*, *18*, 423.
7. Webster, T. B. (1980). *Water Pollution Control*, *79*, 511.
8. Allen, E. A., Bartlett, P. K. N., and Ingram, G. (1984). *Analyst (London)*, *109*, 1075.
9. Dohremann, A. and Kleist, H. (1979). *Analyst (London)*, *104*, 1030.
10. Brooks, R. R., Presley, B. J., and Kaplan, I. R. (1967) *Talanta*, *14*, 809.
11. Kremligg, K. and Peterson, H. (1974). *Analytica Chimica Acta*, *70*, 35.
12. Kinrade, J. D. and Van Loon, B. C. (1974). *Analytical Chemistry*, *46*, 1894.
13. Jan, T. K. and Young, D. R. (1978). *Analytical Chemistry*, *50*, 1250.

14. Stolzberg, R. J. (1975). In *Analytical methods in oceanography*, ed. T. R. P. Gibb, Jr, American Chemical Society, Washington, DC, Advanced Chem. No. 147, p. 30.
15. Danielson, L., Magnusson, B., and Westerlund, S. (1978). *Analytica Chimica Acta*, *98*, 45.
16. Sturgeon, R. E., Berman, S. S., Desauiniers, A., and Russel, D. S. (1980). *Talanta*, *27*, 85.
17. Magnusson, B. and Westerlund, S. (1981). *Analytica Chimica Acta*, *131*, 63.
18. Armansson, H. (1977). *Analytica Chimica Acta*, *88*, 89.
19. Bruland, K. W., Franks, R. P., Knauer, G. A., and Martin, J. H. (1979). *Analytica Chimica Acta*, *105*, 233.
20. Analytical Quantity Control (Harmonized Monitoring) Committee, Water Research Centre, Harlow, Bucks, UK. (1985). *Analyst (London)*, *110*, 109.
21. Bradshaw, S., Gascoigne, A. J., Headbridge, J. B., and Moffett, J. H. (1987). *Analytica Chimica Acta*, *197*, 323.
22. Regan, J. G. T. and Warren, J. (1978). *Analyst (London)*, *103*, 447.
23. Chakraborti, D., Adams, F., Van Mol, W., and Irgolic, K. J. (1987) *Analytica Chimica Acta*, *196*, 23.
24. Cockroft, H. R., Nield, D., and Ramson, L. (1977). *Technical Report TR 59. Atomic Absorption Spectrometric Method for the determination of lead and cadmium in water*, Water Research Centre, Medmenham, UK.
25. British Standards Institution UK (1987). *BS 6068 Section 2.29. Determination of cobalt, nickel, copper, zinc, cadmium in flame atomic absorption spectrometric methods*.
26. Subramanian, K. S., Meranger, J. C., and McCurdy, R. F. (1984). *Atomic Spectrometry*, *5*, 192.
27. Department of the Environment (1976). *Methods for the examination of waters and associated materials. Cadmium in potable water by atomic absorption spectrophotometry. Tentative method*, HMSO, London.
28. Pakalns, P. (1981). *Water Research*, *15*, 7.
29. Department of the Environment (1976). *Methods for the examination of waters and associated materials. Lead in potable waters by atomic absorption spectrophotometry*, HMSO, London.
30. Department of the Environment/National Water Council Standing Committee of Analysts, (1981). *Methods for the examination of waters and associated materials. Nickel in potable waters. Tentative methods*, HMSO, London.
31. Bone, K. M. and Hibbert, W. D. (1979). *Analytica Chimica Acta*, *107*, 219.
32. Childs, E. A. and Gaffke, J. N. (1974) *Journal of Association of Official Analytical Chemists*, *57*, 360
33. Shiraishi, N., Hasegawa, T., Hisayuki, T., and Takahashi, H. (1972). *Japan Analyst*, *21*, 705.
34. Sourova, J. and Capkova, A. (1980). *Vodni. Hospodarstvi, Series B*, *30*, 133.
35. Tweeten, T. N. (1976). *Analytical Chemistry*, *48*, 64.

36. Tao, H., Miyazaki, A., Bansho, K., and Umezaki, Y. (1984). *Analytica Chimica Acta*, *156*, 159.
37. Moore, R. V. (1982) *Analytical Chemistry*, *54*, 895.
38. Sugiyama, M., Fujino, O., Kihara, S. and Matsui, M. (1986). *Analytica Chimica Acta*, *181*, 159.
39. Smith, C. L., Matoaka, J. M., and Willson, W. R. (1984). *Analytical Letters*, *17*, 1715.
40. Sokolovich, V. B., Lel'chuk Yu, L., and Detkova, G. A. (1970). *Izv. Tomsk. Politekh. Inst*, *163*, 130. Ref: *Zhur. Khim.* (1971). 199D, (16). Abstract No. 16G188.
41. Sadilikova, M. (1968) *Mikrochimica Acta*, *5*, 934.
42. Lo, J. M., Wei, J. C., Yang, M. H., and Yeh, S. S. (1982) *Journal of Radioanalytical Chemistry*, *72*, 571.
43. Lo, J. M., Wei, J. C., and Yeh, S. J., (1977). *Analytical Chemistry*, *49*, 1146.
44. Shatipov, E. B. and Khudaibenganov, A. I. (1970). *Izv., Akad. Nauk. Uzbek. S.S.R. Ser. Fiz water Nauk*, (6), 55.
45. Ya, J. C. and Wai, C. M. (1984). *Analytical Chemistry*, *56*, 1689.
46. Lo, J. G. and Yang, J. Y. (1985). *Journal of Radioanalytical and Nuclear Chemistry Letters*, *94*, 311.
47. Marcie, F. J. (1967). *Environmental Science and Technology*, *1*, 164.
48. Wanatabe, H., Berman, S., and Russel, D. S. (1972). *Talanta*, *19*, 1363.
49. Tisue, T., Suls, C., and Keel, R. T. (1985). *Analytical Chemistry*, *57*, 82.
50. Adeljou, S. B. and Brown, K. A. (1987). *Analyst (London)*, *112*, 221.
51. Rigin, V. I. and Yurtaev, P. V. (1986). *Soviet Journal of Water Chemistry and Technology*, *8*, 77.
52. Ihnat, M., Gordon, A. D., Gaynor, L. D., Berman, S. S., Desauliers, A., Stoeppler, M., and Valenta, D. (1980). *International Journal of Environmental Analytical Chemistry*, *8*, 259.
53. Sachdev, S. L. and West, P. W. (1970). *Environmental Science Technology*, *4*, 749.
54. Chormann, F. H., Spencer, M. J., Lyons, W. B., and Mayewski, P. A. (1985). *Chemical Geology*, *53*, 25.
55. Woodriff, R., Culner, B. R., Shrader, D., and Super, A. B. (1973). *Analytical Chemistry*, *45*, 230.
56. Chau, Y. K. and Saitoh, H. (1970) *Environmental Science and Technology*, *4*, 839.
57. Shevchuk, I. A. and Metel, N. I. (1987). *Soviet Journal of Water Chemistry and Technology*, *9*, 74.
58. Petrov, B. I. Oshchepkova, A. P. Zhipovistev, U. P., and Nemkovskii, B. B. (1981). *Soviet Journal of Water Chemistry and Technology*, *3*, 51.
59. Kato, K. (1977). *Talanta*, *24*, 503.
60. Lapid, J., Munster, M. T., Forhi, S., Erni, M., and Kaloucher, L. (1984). *Journal of Radioanalytical and Nuclear Letters*, *86*, 321.
61. Testemale, G. and Leredde, S. L. (1970). In *Report CEA-R-3908*. Centre of Nuclear Studies, Forenay-aux. Roses, France.
62. Shimoishi, Y. and Toei, K. (1978). *Analytica Chimica Acta*, *100*, 65.

63. Uchida, H., Shimoishi, Y., and Toei, K. (1980). *Environmental Science and Technology*, *14*, 541.
64. Saitoh, K., Kabayashi, M. and Suzuki, N. (1981). *Analytical Chemistry*, *53*, 2309.
65. Nakashima, S. and Toei, K. (1968). *Talanta*, *15*, 1475.
66. Gosink, T. A. and Reynolds, P. J. (1975). *Journal of Marine Science Communications*, *1*, 10.
67. Shimoishi, Y. (1973). *Analytica Chimica Acta*, *64*, 465.
68. Young, J. W. and Christian, G. D. (1973). *Analytica Chimica Acta*, *65*, 127.
69. Ross, W. D., Scribner, W. G., and Sievers, R. E. (1970). In *Reprint of 8th Int. Symp on Gas Chromatography*, Ballsridge, Dublin, Ireland. September.
70. Measures, C. I. and Edmond, J. M. (1986). *Analytical Chemistry*, *58*, 2065.
71. Belcher, R., Khalique, A., and Stephen, W. L. (1978). *Analytica Chimica Acta*, *100*, 503.
72. Komarek, J., Horak, J., and Sommer, L. (1974) *Collection Czechoslovakian Chemical Communications*, *39*, 92.
73. Nikolaeva, E. M. (1972). *Trudy Perm. med. Inst*, *108*, 17. Ref: *Zhur Khim.*, (1973) 199D (4) Abstract 4G84.
74. Yotsuyanagi, T., Takeda, Y., Yamashita, R., and Aomura, K. (1973). *Analytica Chimica Acta*, *67*, 297.
75. Fujinaga, T. and Takamatsu, T. (1970). *Journal of Chemical Society of Japan, Pure Chemistry Section*, *91*, 1165.
76. Doolan, K. J. and Smythe, L. E. (1973). *Talanta*, *20*, 241.
77. Gemmer Colos, V., Tuss, H., Saur, D., and Neeb, R., (1981). *Fresenius Zeitschrift fur Analytische Chemie*, *307*, 347.
78. Wenger, R. and Hogel, O. (1971). *Mitt. Geb. Lebensmittelunters u. Hygiene*, *62*, 1.
79. Karrey, J. S. and Goulden, P. D. (1975). *Atomic Absorption Newsletter*, *14*, 33.
80. Chao, T. T. and Ball, J. W. (1971). *Analytica Chimica Acta*, *54*, 166.
81. Andukinova, M. M., Mordberg, G. L., and Nakhorossheva, M. P. (1970). In *The Isolation of Strontium-90. Collection of Radiometric and Gamma Spectrometric Methods of Analysing Materials in the Environment*, Leningrad, p. 26.
82. Tamhina, B., Herak, M. J., and Jayodic, V. (1973). *Craot. Chem. Acta*, *45*, 593.
83. Gorbushina, L. V., Zhil'tsova, L. Y., Matveeva, E. N., Surganova, N. A., Tenyaev, V. G., and Tyminskii, V. G. (1972). *Journal of Radioanalytical Chemistry*, *10*, 165.
84. Garcia-Leon, M., Piazza, C., and Madunga, G. (1984). *International Journal of Applied Radiation and Isotopes*, *35*, 957.
85. Agrawal, Y. K., Chrattopadhyaya, M. C., Abbasi, S. A., and Bodas, M. G. (1973). *Separation Science*, *8*, 613.
86. Agrawal, Y. K. (1973). Separation Science, *8*, 709.
87. Abbasi, S. A. (1988). *International Journal Environmental Analytical Chemistry*, *33*, 113.

88. Kish, P. P. and Zimomrya, I. I. (1969). *USSR Zavod Lab*, *35*, 541.
89. Lazazev, A. I. and Lazareva, V. I. (1969). *Zhur. Analit. Khim*, *24*, 395.
90. Yagodnitsyr, M. A. (1970). *Gig. Savit* (11), 62. Ref: *Zhur Analit Khim*, 19GD (10), (1971) Abstract No. 10G190.
91. Joshi, S. R., Srivanstava, P. K., and Tandon, S. N. (1973). *Journal of Radioanalytical Chemistry*, *13*, 343.
92. Shinde, V. M. and Khopar, S. M. (1969). *Chemia. Analit*, *14*, 749.
93. Chau, Y. K. and Lum Shue Chan, L. (1970). *Analytica Chimica Acta*, *50*, 201.
94. Talmi, Y. and Audrin, A. W. (1974). *Analytical Chemistry*, *46*, 2122.
95. Adam, J. and Pribil, R. (1973). *Talanta*, *20*, 1344.
96. Adeloju, S. B. and Brown, K. A. (1987). *Analyst (London)*, *112*, 221.
97. Alimarin, I. P., Tarasevich, N. I., and Isalev, D. L. (1972). *Zhur Analit. Khim*, *27*, 647.
98. Savitskii, V. N. Paloshenko, V. I. and Osadchii, V. I. (1987). *Journal of Analytical Chemistry of USSR*, *42*, 540.
99. Weber, G. (1986). *Analytica Chimica Acta*, *186*, 49.
100. Golwelker, A., Patel, K. S. and Mishra, R. K. (1988). *International Journal of Environmental Analytical Chemistry*, *33*, 185.
101. Rashid, M. and Ejaz, M. (1986) *Mikrochemica Acta*, No. 3/4 191.
102. Ejaz, M., Zuha, S., Dit, W., Akhtar, A., and Chandri, S. A. (1981) *Talanta*, *28*, 441.
103. Silva, M. and Valcarcel, M. (1982) *Analyst (London)*, *107*, 511.
104. Brueggemeyer, T. W. and Caruso, J. A. (1982). *Analytical Chemistry*, *54*, 872.
105. Ejaz, M., and Qureshi, M. A. (1987), *Talanta*, *34*, 337.
106. Korkisch, J., Sorio, E., and Stelstan, F. (1976). *Talanta*, *23*, 289.
107. Christianson, T. F., Busch, J. E., and Krogh, S. C. (1976). *Analytical Chemistry*, *48*, 1051.
108. Savitsky, V. N., Peleshenko, V. I., and Osadchiy, C. (1986). *Hydrobiological Journal*, *1*, 60.
109. Tervero, M. and Gracia, I. (1983). *Analyst (London)*, *108*, 310.
110. Butler, L. R. P. and Matthews, P. M. (1966). *Analytica Chimica Acta*, *36*, 319.
111. Chau, Y. K. and Lum Shue Chan, K. (1969). *Analytica Chimica Acta*, *48*, 205.
112. Akama, Y., Nakai, T., and Kawamura, F. (1979). *Nippon Kaisui Gakkai-shi*, *33*, 180.
113. Fujinaga, T., Kusaka, Y., Koyama, M., Tsuiji, H., Mitsui, T., Imai, S., Okuda, J., and Takamatsu, T. (1973). *Journal Radioanalytical Chemistry*, *13*, 301.
114. Kulathilake, A. I. and Chatt, A. (1980). *Analytical Chemistry*, *52*, 828.
115. Moore, P. J. (1970). *Transactions of the Institute of Minerals and Metallurgy Section* B, *79*, 107.
116. Chambers, J. C. and McClellan, B. E. (1976). *Analytical Chemistry*, *48*, 2061.
117. Korenaga, T., Motomizu, S., and Toei, K. (1980). *Analyst (London)*, *105*, 328.
118. Kentner, E. and Zeitlin, H. (1970). *Analytica Chimica Acta*, *49*, 587.

119. Motomizu, S. (1973). *Analytica Chimica Acta*, *64*, 217.
120. Forster, W. and Zeitlin, H. (1966). *Analytica Chimica Acta*, *34*, 211.
121. Riley, J. and Topping, G. (1969). *Analytica Chimica Acta*, *44*, 234.
122. Armitage, B. and Zeitlin, H. (1971). *Analytica Chimica Acta*, *53*, 47.
123. Going, J., Wesenberg, G., and Andrejat, G. (1976). *Analytica Chimica Acta*, *81*, 349.
124. Korkisch, J. and Sorio, A. (1975). *Analytica Chimica Acta*, *79*, 207.
125. Gurtler, O. (1977) *Fresenius Zeitschrift fur Analytische Chemie*, *284*, 206.
126. Bruland, K. W., Franks, R. P., Knauer, G. A., and Martin, J. H. (1979). *Analytical Chemistry*, *105*, 233.
127. Boyle, E. A. and Edmond, J. A. (1977). *Analytica Chimica Acta*, *91*, 189.
128. Bruland, K. W. (1980). *Science Letters*, *47*, 176.
129. Boyle, E. A., Huested, S. S., and Jones, S. P. (1981). *Journal of Geographical Research*, *86*, 8048.
130. Kentner, E., Armitage, D. B., and Zeitlin, H. (1969). *Analytica Chimica Actu*, *45*, 343.
131. Yatsimirskii, K. B., Ewel'Yakov, E. M., Pavlova, V. K., and Savichenko, Ya.S. (1979) *Okeanologiya*, IIII, *10*. Ref: *Zhur Khim*. 19GD Abstract No. 11G, 203 (11).
132. Nishimura, M., Matsunaga, K., Kudo, T., and Obara, F. (1973). *Analytica Chimica Acta*, *65*, 446.
133. Yang, C. Y., Shih, J. S., and Yeh, Y. C. (1981). *Analyst (London)*, *106*, 385.
134. Korkisch, J. and Koch, W. (1973). *Mikrochimica Acta*, *1*, 157.
135. Korkisch, J. (1972) *Mikrochimica Acta*, 687.
136. Flynn, W. W. (1973), *Analytica Chimica Acta*, *67*, 129.
137. Brooks, R. R., Presley, B. J., and Kaplan, I. R. (1967). *Talanta*, *14*, 809.
138. Kremling, K. and Peterson, H. (1974). *Analytica Chimica Acta*, *70*, 35.
139. Kinrade, J. D. and Van Loon, J. C. (1974). *Analytical Chemistry*, *46*, 1894.
140. Jan, T. K. and Young, D. R. (1978). *Analytical Chemistry*, *50*, 1250.
141. Stolzberg, R. J. (1975). In *Analytical Methods in Oceanography*, ed. T.R.P. Gibb, Jr, Advanced Chemistry, No. 147, American Chemical Society, Washington DC, p. 30.
142. Danielsson, L., Magnusson, B., and Westerlund, S. (1978). *Analytica Chimica Acta*, *98*, 45.
143. Sturgeon, R. E., Berman, S. S., Desaulniers, A., and Russell, D. S. (1980). *Talanta*, *27*, 85.
144. Magnusson, B. and Westerlund, S. (1981). *Analytica Chimica Acta*, *131*, 63.
145. Armansson, H. (1977). *Analytica Chimica Acta*, *88*, 89.
146. Bruland, K. W., Franks, R. P., Knauer, G. A., and Martin, J. H. (1979). *Analytica Chimica Acta*, *105*, 233.
147. Moore, R. M., Burton, J. D., Williams, P.J.le B., and Young, M. L. (1979). *Cosmochimica Acta*, *43*, 919.
148. Subramanian, K. S. and Meranger, J. C. (1979) *International Journal of Environmental Analytical Chemistry*, *7*, 25.
149. Sugimae, A. (1980). *Analytica Chimica Acta*, *121*, 331.

150. Statham, P. J. (1985). *Analytica Chimica Acta*, *169*, 149.
151. Burton, J. D. (1981). In *Trace Metals in Seawater. Procedures of a NATO Advanced Research Institute on Trace Metals in Seawater*, 30/3–3/4/81. Sicily, Italy, eds C. S. Wong, *et al.*, Plenum Press, New York, p. 419.
152. Brooks, R. R., Presley, B. J., and Kaplan, I. R. (1967). *Talanta*, *14*, 809.
153. Brewer, P. G., Spencer, D. W., and Smith, C. L. (1969). *American Society of Testing Materials*, *443*, 70.
154. Moore, R. M. (1981). In *Trace Metals in Seawater. Proceedings of a NATO Advanced Research Institute on Trace Metals in Seawater*. 30/3–3/4/81, Sicily, Italy, eds C. S. Wong *et al.*, Plenum Press, New York.
155. Danielson, L., Magnusson, G. B., and Westerlund, S. (1978). *Analytica Chimica Acta*, *98*, 47.
156. Apte, S. C. and Gunn, A. M. (1987). *Analytica Chimica Acta*, *193*, 147.
157. Filippelli, M. (1984). *Analyst (London)*, *109*, 515.
158. Lo, J. M., Yu, J. C., Hutchinson, F. I., and Wal, C. M. (1982). *Analytical Chemistry*, *54*, 2536.
159. Sugimae, A. (1980). *Analytica Chimica Acta*, *121*, 331.
160. Muyazaki, A., Kimuka, A., Bansho, K., and Amezaki, Y. (1981). *Analytica Chimica Acta*, *144*, 213.
161. Blockaert, J. A. C., Leis, F., and Laguna, K. (1981). *Talanta*, *28*, 745.
162. Jones, J. S., Harrington, D. E., Leone, B. A., and Branstedt, W. R. (1983). *Atomic Spectroscopy*, *4*, 49.
163. Buchanan, A. S. and Hannaker, P. (1984). *Analytical Chemistry*, *56*, 1379.
164. Van der Berg, C. M. G. (1985). *Marine Chemistry*, *16*, 121.
165. Brugman, L., Magnusson, B., and Westerlund, S. (1983). *Marine Chemistry*, *13*, 327.
166. Shannon, L. L. and Orden, M. J. (1970). *Analytica Chimica Acta*, *52*, 166.
167. Flynn, A. (1970). *Analytical Abstracts*, *18*, 1624.
168. Tseng, C. L., Hsieh, Y. S., and Yong, M. H. (1985). *Journal of Radioanalytical and Nuclear Chemistry Letters*, *95*, 359.
169. Klinkhammer, G. P. (1980). *Analytical Chemistry*, *42*, 117.
170. Hiiro, K., Tanaka, T., and Sawada, T. (1972). *Japan Analyst*, *21*, 635.
171. Moore, R. M. (1981). In *Trace Metals in Seawater, Proceedings of a NATO Advanced Research Institute on Trace Metals in Seawater*, 30/3–3/4/81, Sicily, Italy, eds C. S. Wong *et al.*, Plenum Press, New York.
172. Chau, Y. K. and Lum-Shui-Chan, K. (1969). *Analytica Chimica Acta*, *48*, 205.
173. Landing, W. M. and Bruland, K. W. (1980). *Earth Planet Science Letters*, *49*, 45.
174. Armitage, B. and Zeitlin, H. (1971). *Analytica Chimica Acta*, *53*, 47.
175. Morris, A. W. (1968). *Analytica Chimica Acta*, *42*, 397.
176. Virmani, Y. P. and Zeller, E. J. (1974). *Analytical Chemistry*, *46*, 324.
177. Zharikov, V. F. and Senyavin, M. K. (1970). *Trudy gos okeanogr. Inst*, (101). Ref: *Zhur Khim* 19GD, (7) Abstract No. 7G189.

178. Monieu, H. and Stangel, R. (1982). *Fresenius Zeitschrift für Analytische Chemie*, *311*, 209.
179. Hirao, Y., Fukumoto, K., Sugisaki, H., and Kimura, K. (1979). *Analytical Chemistry*, *51*, 651.
180. Rampon, H. and Cavalier, R. (1972). *Analytica Chimica Acta*, *60*, 226.
181. Bruland, K. W., Franks, R. P., Knauer, G. A., and Martin, J. H. (1979). *Analytica Chimica Acta*, *105*, 233.
182. Tsalev, D. L., Alimarin, I. P., and Neiman, S. I. (1972). *Zhur Analit. Khim*, *27*, 1223.
183. El-Enamy, F. F., Mahmond, K. F., and Varma, M. M. (1979). *Journal of the Water Pollution Control Federation*, *51*, 2545.
184. Pellenberg, R. E. and Church, T. M. (1978). *Analytica Chimica Acta*, *97*, 81.
185. Armansson, H. (1979). *Analytica Chimica Acta*, *110*, 21.
186. Brugmann, L., Danielsson, L. R., Magnusson, B., and Westerlund, S. (1983). *Marine Chemistry*, *13*, 327.
187. Boyle, E. A. and Edmond, J. M. (1977). *Analytica Chimica Acta*, *91*, 189.
188. Jan, T. K. and Young, D. R. (1978). *Analytical Chemistry*, *50*, 1250.
189. Bruland, K. W., Franks, R. B., Knawer, G. A., and Maryin, J. H. (1979). *Analytica Chimica Acta*, *105*, 233.
190. Rasmussen, L. (1981). *Analytica Chimica Acta*, *125*, 117.
191. Ediger, R. D., Peterson, G. E., and Kerber, J. D. (1974). *Atomic Absorption Newsletter*, *13*, 61.
192. Smith, R. G. and Windom, H. L. (1980). *Analytica Chimica Acta*, *113*, 39.
193. Danielsson, L. G., Magnusson, B., and Westerlund, S. (1978). *Analytica Chimica Acta*, *98*, 47.
194. Sturgeon, R. E., Berman, S. S., Desauliniers, A., and Russell, D. S. (1980). *Talanta*, *27*, 85.
195. Sperling, K. R. (1980). *Fresenius Zeitschrift fur Analytische Chemie*, *301*, 294.
196. Sturgeon, R. E., Berman, S. S., Desauliniers, J. A. H., Mykytink, A. P., McLaren, J. W., and Russell, D. M. (1980). *Analytical Chemistry*, *52*, 1585.
197. Sperling, K. R. (1982) *Fresenius Zeitschrift fur Analytische Chemie*, *54*, 2536.
198. Lo, J. M., Hutchinson, J. C., and Wal, C. M. (1982). *Analytical Chemistry*, *54*, 2536.
199. Lo, J. M., Yu, J. C. Hutchinson, F. I., and Wal, C. M. (1982). *Analytical Chemistry*, *54*, 2536.
200. Statham, P. J. (1985). *Analytica Chimica Acta*, *169*, 149.
201. Danielsson, L. G., Magnusson, B., and Westerlund, S. (1982). *Analytica Chimica Acta*, *144*, 183.
202. Sturgeon, R. E., Berman, S. S., and Desauliniers, J. A. (1980). *Analytical Chemistry*, *52*, 1585.
203. Shimoishi, Y. (1973). *Analytica Chimica Acta*, *64*, 465.
204. Measures, C. I. and Burton, J. D. (1980). *Analytica Chimica Acta*, *120*, 177.
205. Sui, K. W. M. and Berman, S. S. (1984). *Analytical Chemistry*, *56*, 1806.

206. Lee, M. L. and Burrell, D. C. (1973). *Analytica Chimica Acta*, *66*, 245.
207. Siu, W. M., Bednas, H. E., and Berman, S. S. (1983). *Analytical Chemistry*, *55*, 473.
208. Heumann, K. G. (1980). *Toxicological Environmental Chemical Review*, *3*, 111.
209. Colby, B. N., Rosecrance, A. E., and Colby, M. E. (1981). *Analytical Chemistry*, *53*, 1907.
210. Mykytiuk, A. P., Russell, D. S., and Sturgeon, R. E. (1980). *Analytical Chemistry*, *52*, 1281.
211. Sturgeon, R. E., Berman, S. S., Willie, S. N., and Desauiniers, J. A. H. (1981). *Analytical Chemistry*, *53*, 2337.
212. Lee, M. G. and Burrell, D. C. (1972). *Analytica Chimica Acta*, *62*, 153.
213. Parker, C. R. (1972). In *Water Analysis by Atomic Absorption Spectroscopy*, Varian Techtron.
214. Clem, R. G. and Hodgson, A. T. (1978). *Analytical Chemistry*, *50*, 102.
215. Wilson, H. N. (1951). *Analyst (London)*, *76*, 65.
216. Wilson, H. N. (1954). *Analyst (London)*, *79*, 535.
217. MacDonald, A. M. G. and Rivero, A. M. (1967). *Analytica Chimica Acta*, *37*, 414.
218. Kirkbright, G. F., Narayanaswamy, R., and West, T. S. (1972). *Analyst (London)*, *97*, 174.
219. Babko, Yu.F. Shkaravskii, F., and Ivanshkovich, E. M. (1967). *Ukranian Chemical Journal*, *33*, 30.
220. Pilipenko, A. T., and Shkaravskii, Yu.F., (1974). *Zhur Anal. Khim*, *29*, 716.
221. Shida, J. and Matsuo, T. (1980). *Bulletin of the Chemical Society of Japan*, *53*, 2868.
222. Ducret, L. and Drouillas, M. (1959). *Analytica Chimica Acta*, *21*, 86.
223. Sudakov, F. P., Klitina, V. I. Ya, T., and Dan'shova, H. (1966). *Zhur Anal. Khim*, *21*, 1333.
224. Babko, A. K., Shkaravskii, Yu.F., and Kulik, V. I. (1966). *Zhur Anal. Khim*, *21*, 196.
225. Matsuo, T., Shida, J., and Kurihara, W. (1977). *Analytica Chimica Acta*, *91*, 385.
226. Kirkbright, G. F., Narayanasamy, R., and West, T. S. (1971). *Analytical Chemistry*, *43*, 1434.
227. Motomizu, S., Wakimoto, T., and Toei, K. (1982). *Analytica Chimica Acta*, *138*, 329.
228. Motomizu, S., Wakimoto, T., and Toei, K. (1984). *Talanta*, *31*, 235.
229. Motomizu, S. and Oshima, M. (1987) *Analyst (London)*, *112*, 295.
230. Chambe, A. and Gupta, V. K. (1983) *Analyst (London)*, *108*, 1141.
231. Fogg, A. G. and Yoo, K. S. (1976). *Analytical Letters*, *9*, 1035.
232. Muyazaki, A., Kimura, A., and Umezaki, Y. (1981). *Analytica Chimica Acta*, *127*, 93.
233. Muyazaki, A., Kimura, A., and Umezaki, Y. (1982). *Analytica Chimica Acta*, *138*, 121.
234. Bet-Pera, F., Srivasstava, A. K., and Jaselskis, B. (1981). *Analytical Chemistry*, *53*, 561.

235. Pavlova, M. W., Podal'skaya, B. L., and Shafran, I. G. (1972). *Trudy uses Naucho issled. Inst. Khim. Reakt. osobo Christ. Khim. Veshchesto*, *34*, 185. Ref: Zhur. Khim. 19GD, (1973) (11) Abstract No. 11G200.
236. Fukamachi, K., Morimoto, M., and Yanagawa, M. (1972). *Japan Analyst*, *21*, 26.
237. Yanagisawa, M., Suzuki, M., and Takequichi, T. (1973). *Mikrochimica Acta*, *3*, 475.
238. Subramanian, K. S. (1988). *Analytical Chemistry*, *60*, 11.
239. Schaller, H. and Neeb, R. (1987). *Fresenius Zeitschrift fur Analytische Chemie*, *327*, 170.
240. Miyazaki, A. and Brancho, K. (1987). *Analytica Chimica Acta*, *198*, 297.
241. Miyazaki, A. and Brancho, K. (1987). *Spectrochimica Acta*, *423*, 227.
242. Chakraborty, D. and Das, A. K. (1988). *Atomic Spectroscopy*, *9*, 115.
243. Metzger, M. and Braun, H. (1986). *Analytica Chimica Acta*, *189*, 263.
244. Mok, W. M. and Wai, C. M. (1987). *Analytical Chemistry*, *59*, 233.
245. Mok, W. M., Shah, N. W., and Wai, C. M. (1986). *Analytical Chemistry*, *58*, 110.
246. Stary, J., Zeman, A., Kratzer, K. and Prasilova, J. (1980). *International Journal of Environmental Analytical Chemistry*, *8*, 49.
247. Dix, K., Cappon, C. J., and Toribara, T. Y. (1987). *Journal of Chromatographic Science*, *25*, 164.
248. Nasu, T. and Kau, M. (1988). *Analyst (London)*, *113*, 1685.
249. Batley, G. E., Florence, T. M., and Kennedy, J. R. (1973). *Talanta*, *20*, 987.
250. Chau, Y. K., Wong, P. T. S., and Kramer, O. (1983). *Analytica Chimica Acta*, *146*, 211.
251. Chau, Y. K., Wong, P. T. S., and Goulden, P. D. (1976). *Analytica Chimica Acta*, *85*, 421.
252. Schintu, M., Kauri, T, Contu, A., and Kudo, A. (1987). *Ecotoxicology and Environmental Safety*, *14*, 208.
253. Filippi, M. (1984). *Analyst (London)*, *109*, 515.
254. Meinema, H. A., Burger, N., and Wiersina, I. (1978). *Environmental Science and Technology*, *12*, 288.
255. Boyle, E. A., Handy, B., and Van Green, A. (1987). *Analytical Chemistry*, *59*, 1499.
256. Motomizu, S., Onada, M., Oshima, M., and Iwachido, T. (1988). *Analyst (London)*, *113*, 743.
257. Atallah, R. H., Christian, G. D., and Hartenstein, S. D. (1988). *Analyst (London)*, *113*, 463.
258. Murata, M., Omatsu, M., and Muskimoto, S. (1984). *X-Ray Spectrometry*, *13*, 83.

3
MACRORETICULAR NON-POLAR RESINS

3.1 Organics

The principle of adsorption techniques for preconcentrating samples is quite simple. A large volume of sample is contacted with or passed down a column of a solid material which removes organic material from the water into the solid. Many mechanisms can be responsible for this; adsorption phenomenon is one. In the next stage, the organics (and in a limited number of cases cations or organometallic compounds, see Sections 3.2 and 3.3) are removed from the solid into a volume of a solvent or a chemical reagent which is relatively low compared to the original volume of sample taken, and subsequently examined by gas chromatography or high performance liquid chromatography either of which might be coupled to a mass spectrometer. Alternatively, the adsorbent might be heated to release the organics which are then swept into a cryogenic trap for subsequent analysis by the aforementioned chromatographic techniques or swept directly into a gas chromatograph. In favourable circumstances preconcentration factors of up to 10^4 or even higher are commonly achieved by these techniques. Naturally, the types of resins from which preconcentrated substances can be subsequently released either by solvent extraction or thermal desorption must themselves have no polarity. Only the non-polar resins are discussed in this chapter. Polar (i.e. ion exchange) resins are discussed in Chapters 4 and 5 (Volume 1). Non-polar resins, as opposed to polar resins operate by forming various kinds of loose physical bonds (rather than chemical bonds) with the adsorbed substance, or, in the case of larger adsorbed molecules, may operate on the principle of acting as microfilters, i.e. the pore sizes in the resin are of molecular dimensions.

The first non-polar macroreticular adsorbent to be manufactured was Amberlite XAD-1. Since then, improved non-polar resins (XAD-2 and XAD-4) have become available and the applications of these are discussed below in Section 3.1.1. Tenax G C is a further very popular non-polar macroreticular resin. It is based on 2,6-diphenyl-*p*-phenylene oxide.

3.1.1 XAD-2 and XAD-4 non-polar resins

Non-saline waters—general discussion Daignault *et al.*[1] reviewed the literature on the use of XAD resins to concentrate organic compounds in water including the occurrence of artifacts released into water from these resins; methods of cleaning the resin and/or solvent before use; factors affecting adsorption on the resin (pH value, ionic strength, flow rate, resin capacity and bead size, and presence of humic substances); eluants used and compound recoveries; and resin regeneration. Wigilius *et al.*[2] discussed a systematic approach to adsorption on XAD-2 resin for the preconcentration and analysis of organics in water below the pg litre^{-1} level. The effects of solvent changes on the adsorption of trace organics from water by XAD-2 resin were investigated. Methanol slurried resin was packed into a column and washed with methanol and diethyl ether. Tap water was passed through the column to condition the resin before passing 10–20 litres of an aqueous sample containing 32 model compounds. Tap water or river water containing 200 ng litre^{-1} of each compound was used. The effects of changing the solvent sequence to methanol/diethyl ether/water and to methanol/acetone/water were investigated. Diethyl ether was used to desorb the organics. For a large number of non-polar compounds, recovery was limited by evaporation losses. For polar compounds (alcohols and phenols), adsorption efficiency was critical. Desorption losses were made negligible, except for naphthalene and anthracene. Humic acids gave no decrease in recovery of low molecular weight compounds and no increased breakthrough. The limit of detection was below 20 ng litre^{-1} for a 10 litre sample volume. Acceptable blanks were achieved.

Dietrich *et al.*[3] have made an intercomparison study of the applications of adsorption on XAD resins and methylene dichloride liquid–liquid extraction from the preconcentration of numerous organic compounds from river water. Capillary gas chromatography–mass spectrometry was used to detect and identify compounds. Of the 48 individual chemicals identified over 13 months the most frequently observed were atrazine, and methyl atraton, dimethyl dioxan, 1,2,4-trichlorobenzene, tributylphosphate, triethylphosphate, trimethylindolinone, and the three isomers of tris(chloropropyl) phosphate. Both extraction methods gave similar results in the ng litre^{-1} range but the XAD resin method showed more artefacts. Stephan and Smith[4] showed that adsorption on macroreticular resins is a suitable method for extracting trace organic pollutants from water prior to determination by gas chromatography. The flow rate and pH value of the sample should be controlled to obtain the best results. Extraction efficiencies obtained for various compounds on XAD-2 and XAD-7 resins (i.e. non-polar and mildly polar) are in the range 27–93 per cent (Table 50).

Chang and Fritz[5] pointed out that the XAD-2 resin sorption method was a reliable and thoroughly tested method for determining trace organic pollutants in potable water. However, the solvent evaporation step in this procedure results in a partial or complete loss of volatile

Table 50 Effect of polarity of resin on extraction efficiency of polar and non-polar compounds.

Compound	Extraction efficiency (Averages of 5 extractions) (±2 per cent)	
	XAD-2 (per cent)	XAD-7 (per cent)
Cumene	67	67
Ethyl benzene	60	59
Naphthalene	90	93
n-Hexane	82	83
Phenol	27	45
Octanoic acid	58	81
o-Cresol	59	67
Chlorophenol	70	85

Sample flow: 20 ml min^{-1}; pH: 5.7; concentration 10 ppm.
From Stephan and Smith[4] with permission.

compounds. To avoid these drawbacks a procedure was developed by them in which organic compounds that are sorbed on a resin are thermally desorbed directly on to a gas chromatographic column for analysis. They desorbed the organics on to a Tenax tube from a tube containing XAD-2. This eliminates virtually all of the water entrained in the XAD-2 tube. Then the organics are thermally desorbed from the Tenax tube directly on to the gas chromatographic column. The method is as follows:

Resins:

XAD-2 resin (Rohm and Hass) ground and sieved in the dry state Tenax GC, 60–80 mesh, used without further purification.

Apparatus and equipment:

Hewlett Packard 5756 B gas chromatograph equipped with a linear temperature programmer and FID detector or equivalent. Stainless steel tubing for all columns and injection port liners. The injection port was enlarged by drilling to accommodate the glass sorption tubes. The column for separations was either 6 ft × 1/8 in o.d. packed with 5 per cent OV-17 on Chromosorb W AW DMCS (80–100 mesh) or 4 ft × 1/8 in packed with Tenax GC (60–80 mesh).

Sorption tubes were 2 mm i.d., 10 cm in length. They were filled with 7 cm of either XAD-2 or Tenax GC, held in place with a plug of glass wool at either end. The XAD 2 tubes were conditioned by thermal desorption at 240 °C and repetitive blanks were run until a low, tolerable, background was achieved. A blank run involved wetting the

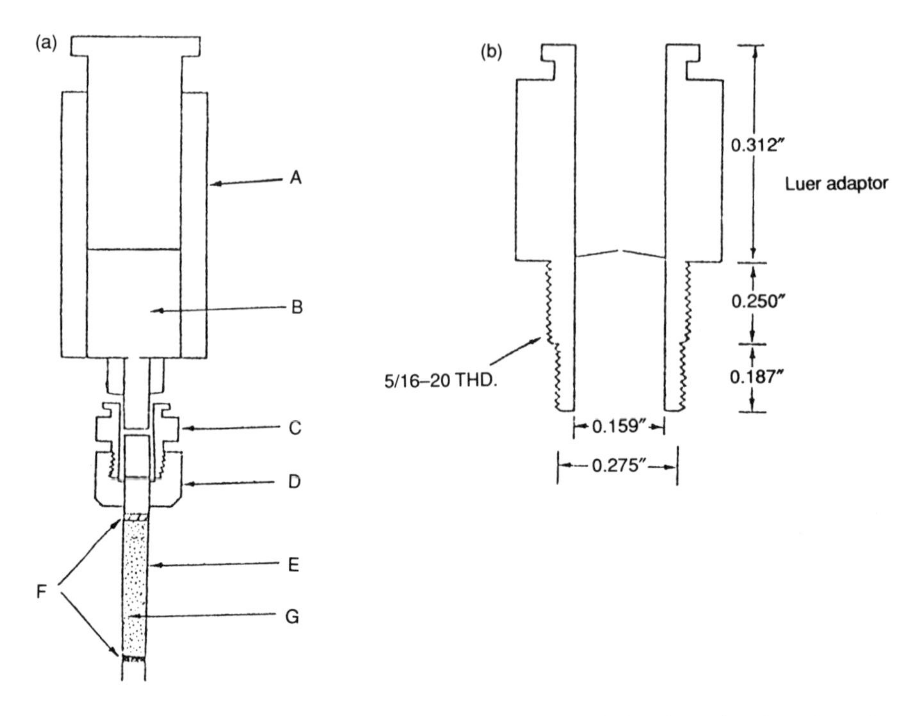

Fig. 13 (a) Minisampler for small water volumes: A, 20 ml glass hypodermic syringe; B, water sample; C, coupler for attaching minicolumn (E) to the syringe; D, 1/4 in Swagelok nut; E, 2 mm i.d. Pyrex tube; F, glass wool plugs; G, 80 mg of 60–80 mesh resin (b) Kel–F–Coupler. From Stephan and Smith[4] with permission.

resin with triply distilled water and following steps 2 and 3 of the analytical procedure. The Tenax GC tube required only a simple conditioning by heating for 1 h at 275 °C to achieve a tolerable blank level.

A minisampler for water in constructed from a 20 ml glass syringe connected to an XAD-2 sorption tube (Fig. 13(a)). The connection is made with the special Kel-F coupler shown in Fig. 13(b).

Procedure:

1. *Sampling*: Pass a 15 ml (or larger) sample through the XAD-2 sorption tube by using a minisampler with hand pressure to force the water through (about 1–2 min). Then force as much residual water as possible from the tube, using 20 ml of air. Cap the sorption tube at both ends unless the analysis is to continue without delay.
2. *Thermal desorption*: Connect the XAD-2 tube to a Tenax sorption tube. Place the XAD-2 tube in a heated zone, maintained at 230 °C with helium flowing at 50 ml min^{-1} into the XAD-2 tube and out of the Tenax tube (the Tenax tube is held at approximately 45°).

Continue desorption from the XAD-2 on to the Tenax tube for 15 min or until the Tenax is visibly dry.

3. *Separation*: Disconnect the two tubes and place the Tenax tube in the modified GC injection port, held at 220 °C and apply a helium carrier gas flow of 20 ml min^{-1}, (the oven section of the chromatograph is at approximately room temperature). Allow 10 min for transfer of organic solutes from the Tenax tube to the column, then separate by raising the temperature at 20 °C min^{-1} up to 200 °C finally holding the temperature for 10 min.
4. *Measurement*: Inject 2–5 ul of standard organic solution in methanol through a zone heated to 220 °C on to a Tenax tube at room temperature (or slightly above). Desorb and chromatograph the organic compounds as in step 3. Compare the heights of the sample and standard peaks to calculate the amounts of sample constituents.

Recoveries obtained by this procedure are tabulated in Table 51. The recoveries of carboxylic acids were not reproducible. This is believed

Table 51 Overall recovery efficiency of resin extraction and thermal desorption method of analysis on XAD-2 for organics in water at the 3–100 μg $litre^{-1}$ level.

Compounds	Recovery efficiency (per cent)	Compounds	Recovery efficiency (per cent)
Alkanes		*Polynuclear aromatics*	
Octane	88	Naphthalene	98
Heptane	81	2-Methylnaphthalene	97
Tridecane	90	*Chlorobenzenes*	
Alkylbenzenes		Chlorobenzene	90
Benzene	90	*o*-Dichlorobenzene	82
Toluene	97	*Ketones*	
o-Xylene	90	Acetone	55
Cumene	82	2-Octanone	83
Ethers		Undecanone	86
Hexyl	80	Acetophenone	92
Benzyl	70	*Alcohols*	
Esters		1-Butanol	<40
Benzyl acetate	95	1-Pentanol	<40
Methyl decanoate	88	1-Octanol	85
Methyl hexanoate	86	1-Decanol	84
Haloforms[a]		*Phenols and acids*	
Chloroform	87	No quantitative results	
Bromodichloromethane	95		

[a] Chlorinated methanes were tested at a concentration of 200 μg $litre^{-1}$ in water.
From Stephan and Smith[4] with permission.

to be due to the difficulty of direct determination of acids by gas chromatography. A selective method for concentration of the acids with subsequent gas chromatography of their derivatives would be more desirable. The recoveries of phenols varied from 4–90 per cent. These inconsistent results are probably due to the high affinity of Tenax GC for phenols. All attempts to improve the yield for phenols failed and the method is not satisfactory for these compounds.

Ryan and Fritz[6] also used thermal desorption of the organics on the resin directly into the gas chromatograph. In this method a small water sample (20–250 ml) is passed through a small tube containing XAD-4 resin; this effectively retains the organic impurities present in the water. This tube is connected to the permanent apparatus and the sorbed organics are thermally transferred to a small Tenex pre-column while the water vapour is vented. The pre-column is closed off, preheated to 275–280 °C and then a valve is opened to plug inject the vaporized sample into a gas chromatograph.

Adsorption tubes:

The tube is made from standard Pyrex glass tubing 8 cm × 2 mm i.d. and filled with a bed of XAD-4 resin, 120–140 mesh, about 2.5 cm in length. The resin is held in place approximately at the centre of the tube by plugs of silanized glass wool. Connections were made with 1/4″ Swagelok nuts and PTFE reducing ferrules. A pair of each were dedicated to the absorption tube.

Purification of XAD-4 resin:

First, pass 5 ml of distilled water through the XAD-4 absorption tube, and then place the tube in a desorption chamber held at 200 °C. During this time, pass helium through the tube at about 20 ml min^{-1}. After about 4 min, raise the temperature to 240 °C for 15 to 20 min. Repeat the entire procedure three more times. This purification method resulted in a tolerably low blank when the tube was tested in blank thermal desorption runs.

Thermal desorption apparatus:

The apparatus is shown in Fig. 14 and consists of the following components:

A. An alluminium block, movable by sliding on its mount. (a) Thermocouple connected to external pyrometer; (b) one 200 watt cartridge heater, controlled by variable transformer; (c) XAD-4 minicolumn.

B. Heated zone insulated with Temp-mat Glass Insulation (courtesy of Pittsburgh Corning, Pittsburgh, Pennsylvania). (d) Two zero-volume high temperature valves, one 4 port and one 6 port as shown (Valco Valve Co., Houston, Texas); (e) high temperature heating

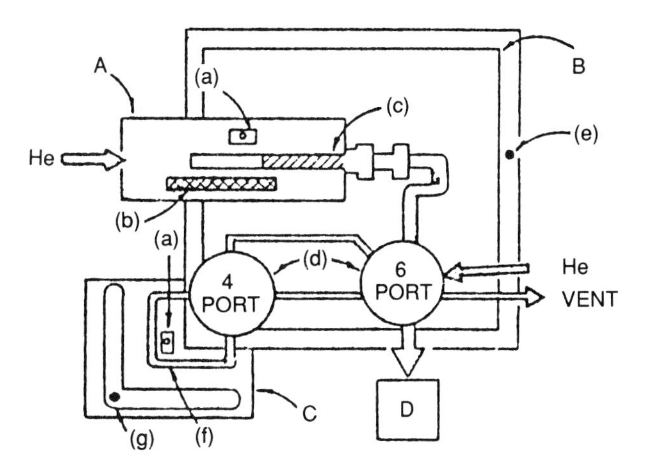

Fig. 14 Thermal desorption apparatus. From Ryan *et al.*[6] with permission.

cord, 200 W (Class Col. Apparatus, Co., Terre Haute, Indiana) controlled by variable transformer.

C. A removable insulated (as B) sheet metal enclosure containing the Tenax desorption heating zone. (f) Tenax precolumn (stainless steel 1/8″ × 18 cm, Tenax GC 80–100 mesh); (g) high temperature heating cord, 200 W (as 3) controlled by temperature controller (Ames Lab, Instrument Group).

D. Gas chromatography (Tracer 560).

Gas chromatography:

Carry out separations on glass columns 6 ft × 2 mm i.d. 10 per cent Carbowax 20 M on Chromosorb W 80–100, 5 per cent FFAP on Chromosorb W 80–100, and 5 per cent OV-1 on Chromosorb W 80–100 stationary phase were used. Run temperature programs at 10 °C min^{-1} with initial temperature ranging from 50–75 °C. Final temperature was usually 190 °C. The gas chromatograph was equipped with dual FID detectors. The attenuation was 4 × 10. This was ideal for samples ranging from 100–300 ng of individual organic impurities.

Thermal desorption procedure:

Connect the minicolumn to the desorption apparatus by means of the two (Swagelok nut) PTFE ferrule connections on the XAD-4 tube. Then thermally desorb the organics from the XAD-4. Typical conditions are: temperature 180–200 °C; time, 10 min; gas flow, helium, 5 ml min^{-1}. The vapour passes through a heated (200–220 °C) zone to a Tenax precolumn (temp. 45 °C). Water passes through the Tenax to vent while the organics are retained.

Close the Tenax precolumn off by means of a 4-port valve equipped with zero volume fittings, and heat to 275–280 °C. Then open the valve to inject the vapour phase into the gas chromatography by diverting the carrier gas to back-flush through the hot precolumn. The total volume of the precolumn, including Tenax volume, is 1.0 ml so the void volume is somewhat less than 1.0 ml.

Open the Tenas desorption chamber while the sample is being chromatographed. This cools the precolumn in preparation for the next sample. During the chromatographic run the desorption block reaches 230–240 °C effectively regenerating the minicolumn.

On completion of the run, remove the XAD-4 tube and place it on a stainless steel sheet which serves as a heat sink. This rapidly cools the XAD-4 tube for the next analysis and minimizes oxygenation of the resin.

The thermal desorption procedure was tested by analysing water samples, each containing a known concentration of a model organic compound. The model compounds were selected to include different organic functional groups and compounds of varying volatility. The precentage recovery for each model compound was calculated by comparing the peak height for the compound in the sample with the peak height of the same compound in a methanol or acetone standard injected directly on to an XAD-4 tube. The recoveries reported in Table 52 are generally quite good at both 10 and 1 μg litre^{-1} levels in water although there is some apparent loss of the halocarbons tested at the lower concentration level.

The recovery (as defined in Table 52) of high boiling compounds was found to be dependent on desorption time and temperature. The reason for this is not entirely clear. Methyl nonyl ketone, for example, requires a higher temperature than normal for efficient desorption. Perhaps the methanol or acetone in the standard is caught by the XAD-4 and helps sweep out the methyl nonyl ketone (the sample has almost no methanol or acetone, so this effect is absent); or the methyl nonyl ketone may be dispersed in a broader band on the XAD-4 tube when taken up from a water sample and a higher temperature is needed to desorb it in a back-flush mode. The higher temperature (225 °C) required for thermal desorption of methyl nonyl ketone greatly enhances the resin blank. However even under these strenuous conditions a 1.0 μg litre^{-1} solution of this ketone can be observed with a signal to noise ratio of better than 10. Volatile ketones give low recoveries at the higher temperature (225 °C) when compared to a standard desorbed at 180 °C. Of course the standard and sample should be desorbed under similar conditions but this comparison does indicate some loss of more volatile compounds if the desorption time and temperature are too great. The partial loss

Table 52 Recovery of organic compounds by thermal desorption from XAD-4 resin.

Compound	Recovery (per cent)[a]		Desorption	
	10 μg litre^{-1} 20 ml	1 μg litre^{-1} 200 ml	Time (min)	Temp (°C)
Toluene	88	90	15	210
Ethylbenzene	79	79	15	210
Indene	96	102	10	180–200[b]
Naphthalene	90	81	15	210
l-Methylnaphthalene	95	97	10	180–200
Hexane	88	65	8	175
Chloroform	93	56	4	175
Dibromomethane	88	57	4	175
Cyclohexanol	98	90	5	200
n-Heptyl alcohol	100	98	6	200
Benzyl alcohol	83	54	13	220
Methyl lsobutyl ketone	100	98	8	200
Amyl lsopropyl ketone	102	99	8	200
Methyl nonyl ketone	96	92	13	220
p-Methyl acetophenone	99	104	13	220
Ethyl heptanoate	96	68	10	200
Octyl acetate	61	32	10	200
Bromobenzene	106	–	10	200
o-Dichlorobenzene	102	–	10	200

[a] The recovery of compounds from water by XAD-4 is based on the comparison of spiked water sample determinations with standards injected directly onto an XAD-4 minicolumn using the 'wet' column procedure. Standards were $1 \times 10^{-17}\ g^{-1}\ \mu l$ in methanol or acetone. Spiked water samples contained 2 μl of the standard solution.
[b] The two numbers represent the initial and final temperatures of the desorption block. This crude temperature programming procedure seemed to minimize resin bleed while maintaining good desorption efficiency.
From Ryan and Fritz[6] with permission.

of highly volatile compounds to vent is a consequence of keeping the Tenax column at a slightly elevated temperature. However, the Tenax must be heated slightly in order to ensure efficient passage of water vapour.

The efficiency of the thermal desorption procedure was determined for several of the model compounds. This was done by comparing the

Table 53 Thermal desorption efficiency.[a]

Compound	Mini column condition	Desorption time (min)	Desorption temperature (°C)	Boiling point (°C)	Fractional[b] recovery
Cyclohexanol	Dry	10	180–200	161	0.94
Cyclohexanol	Wet[a]	10	180–200	161	1.02
Bromobenzene	Wet	8	200–215	156	0.97
Bromobenzene	Wet	10	180–200	156	0.97
Undecanone	Wet	10	210–225	232	0.66
Undecanone	Wet	10	180–200	232	0.50
Chloroform	Wet	10	180–185	62	0.93
Indene	Wet	10	180–200	183	0.99
l-Methylnaphthalene	Wet	13	200–235	245	0.96
l-Methylnaphthalene	Wet	10	180–200	245	0.86

[a] Wet columns are wetted after the sample has been injected into them.
[b] Fractional recovery is the ratio of peak height of injection onto the XAD-4 minicolumn to the peak height of direct injection into the gas chromatograph. Samples were 2×10^{-7} g.
From Ryan and Fritz[6] with permission.

peaks of compounds in standards taken through the entire procedure with peaks of the same compounds in standard injected directly into the gas chromatograph. The data in Table 53 show that except for methyl nonyl ketone the recovery efficiencies of all compounds tested are quite good. Thus for quantitative analysis of actual water samples, the results could be compared with standards injected directly into the gas chromatograph (as in Table 53) or with water standards passed through the XAD-4 tube.

Tateda and Fritz[7] used solvent extraction with carbon disulphide or acetone, rather than a thermal method to desorb adsorbed organics from XAD resins. These workers used a minicolumn 12–1.8 mm × 25 mm containing XAD-4 resin or Spherocarb to adsorb organic contaminants from a 50–100 ml potable water sample. The sorbed organics are eluted by 50–100 ul of an organic solvent and the organic solutes separated by gas chromatography. The procedure is simple, it requires no evaporation step and gives excellent recoveries of model organic compounds added to water. The average recovery was 89 per cent at 2–10 μg litre^{-1} levels and 83 per cent at the 100 μg litre^{-1} level. The average standard deviation is 6.3 per cent. Errors other than sorption and desorption would be included, like evaporation losses from the sample, decomposition, sorption on glassware, calibration of eluate volume, and errors in the GC determination.

Tateda and Fritz[7] compared the adsorbtive properties of XAD-4 resin with those of a spherical carbon molecular sieve of large surface area (Spherocarb). It was found that 100 μg carbon disulphide will elute most organic compounds tested with Spherocarb except for phenols and some strongly sorbed compounds like naphthalene. Spherocarb has one major advantage over XAD-4 for analytical use; stronger retention of low molecular weight polar organic compounds. The minicolumn method while directly applicable for waste water analysis where organics contents are relatively high, does not have the sensitivity needed for the analysis of organics in most potable water samples. Large scale desorption methods such as that described by Junk *et al.*[8] have better sensitivity but suffer from a small fraction of the sample being used for the gas chromatograhic analysis (typically a 2 μl aliquot of 1000 μl of diethyl ether extract). Tateda and Fritz[7] combined their procedure with that of Junk *et al.*[8] by adding 1 ml of the ether concentrate from the larger scale procedure to 50 ml of pure water and then proceeding according to the minicolumn procedure. A large fraction of the sample is thus taken for analysis (2 μl of the 100 μl carbon disulphide eluate) so that the original ether concentrate is further concentrated by a factor of 10. By the standard procedure only small peaks are obtained, Spherocarb shows much larger peaks at the same attenuation (10 × 8). With XAD-4 most peaks are close to the expected 10-fold increase in peak height.

The sensitivity of the minicolumn method for most organic compounds is about 2 μg litre^{-1} which is quite adequate for analysis of waste water and badly contaminated potable water. The major components in a well water sample were indene, methyl indene, methylnaphthalene, acenaphthalene, and acenaphthene. The other peaks were not identified. Using naphthalene as a standard the total concentration of the five major peaks in the well water was estimated to be 260 μg litre^{-1}. The total concentration of all peaks was roughly 325 μg litre^{-1}.

Workers at the Water Research Centre, UK[9] passed up to 5 litres of potable water samples through a column of XAD-2 resin and removed the adsorbed organics with 40 μg of diethyl ether. The ether extract is then concentrated 2500 times and examined by gas chromatography–mass spectrometry. Several hundred organic compounds were examined in this survey, many of which were identified in potable water samples.

The sample (5 litres) was spiked in a precleaned ground glass stopper bottle with a solution (5 μl) of the internal standards in acetone (100 mg litre^{-1}) and passed (40 ml min^{-1}) through a column (13 cm × 1.2 cm i.d.) containing the XAD-2 resin (bed height 6 cm). Deuterated internal standards were used to provide, if necessary, quantitative information on some of the compounds identified. The standards (chlorobenzene-d_5, *p*-xylene-d_{10}, phenol-d_5, naphthalene-d_8, hexadecane-d_{34}, and phenanthrene-d_{10}) were added to each sample immediately before extraction. The resin required extensive purification before use. This involved an initial washing with water, heat desorption (200 °C for 16 h) under a flow (100 ml min^{-1}) of purified nitrogen, and a soxhlet extraction (6 h with methanol). After the sample had passed through the resin, purified diethyl ether (15 ml) was added to the column head and allowed to flow through the column until solvent first appeared at the botton of the column. The flow of ether was stopped and the ether allowed to remain in contact with the resin for 10 min. A further aliquot of ether (10 ml) was added to the column, and all of the ether allowed to flow through into a collection flask. This extract was dried by storing overnight in a freezer (−18 °C) and decanting the ethereal solution off any ice formed. The extract was concentrated (to 250 μl) by evaporation in a Junk vessel fitted with a three-ball Snyder column (Fig. 15), and to the final volume (100 μg) by evaporation using a stream of dry, purified nitrogen. The extract was then suitable for gas chromatography–mass spectrometry.

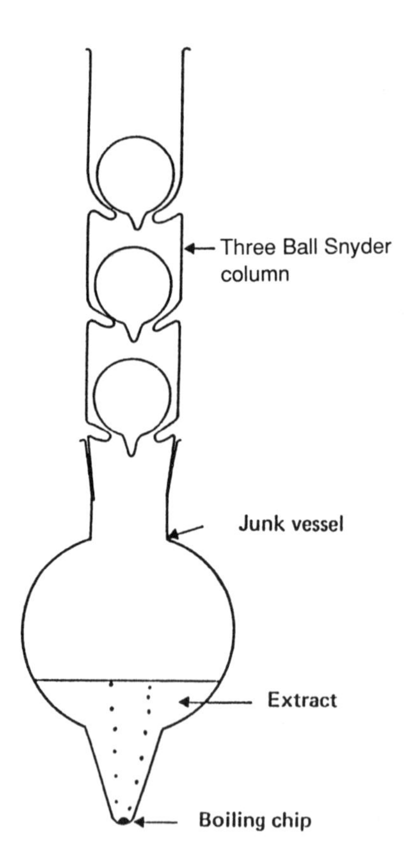

Fig. 15 Apparatus for concentration of XAD-2 resin diethyl ether extract. From Fielding *et al.*[9] with permission.

Stephan *et al.*[10] have described an apparatus for on-site extraction of organic compounds from natural water samples on to XAD-2, XAD-7, and Tenax GC resins. The sampler consist of two peristaltic pumps driven by a battery powered motor. The maximum lift of the peristaltic pumps is approximately 10. Hence the sampler may be used to sample ground water, provided the water table is no more than 10 m below the surface. Operation of the sampler is controlled by an electronic timer. A voltage stabilization circuit is used to control the pumping speed. The rate of sampling may be varied continously between 0 and 60 ml min^{-1} by variation of the pump speed. The electronic timer is designed to allow continuous or automatic intermittent sampling. In the continuous mode, at a sample flow of 25 ml min^{-1} and 1 m lift, the maximum sampling time is approximately 14 h. In the intermittent mode, with timer set to one hour off–one hour on, with a sample flow of 25 ml min^{-1} and 1 m lift, the maximum sampling time is approximately 30 h. Organics were later desorbed in the laboratory from the resin filters in the field apparatus using diethyl ether and methanol and examined by gas chromatography. Very similar results were obtained by the field method and conventional laboratory procedures.

Applications of XAD resins

Polyaromatic hydrocarbons Benoit *et al.*[11] have investigated the use of macroreticular resins, particularly Amberlite XAD-2 resin, in the preconcentration of Ottawa potable water samples prior to the determination of 50 different polyaromatic hydrocarbons by gas chromatography–mass spectrometry.

Water samples were prepared as follows: sampling cartridges, containing 15 g Amberlite XAD-2 macroreticular resin that had been previously cleaned by the method of McNeil *et al.*,[12] were rinsed with 250 ml acetone and washed with at least 1 litre of purified water. The cartridges were attached to a potable water supply and the flow of water was controlled at *c.* 70 ml min^{-1}. When 300 ml of water had been passed through the cartridge, the cartridge was disconnected from the tap and as much water as possible was removed from the cartridge by careful draining followed by the application of vacuum from a water aspirator. The XAD-2 resin was eluted with 300 ml of 15:85 v/v acetone: hexane solution at a flow rate of *c.* 5 ml min^{-1} (all solvents were of 'distilled in glass' quality and were redistilled in an all glass system). The organic layer was dried by passage through a drying column containing anhydrous sodium sulphate over a glass wool plug. Both the sodium sulphate and the glass wool plug were cleaned by successive washings with methylene chloride, acetone, and hexane prior to use. the dried solution was concentrated to a volume of *c.* 3 ml using a rotary evaporator, then quantitatively transferred with acetone to a graduated vial and was further concentrated, using a gentle stream of dry nitrogen gas, to a final volume of 1 ml.

To analyse the solvent extracts a 10 μl aliquot of the concentrated extract was injected into a Finnigan 4000 gas chromatography–mass spectrometry instrument coupled to a 5110 data system. A 3 per cent OV-17 provided the best separation of the detectable polyaromatic hydrocarbons. A 1.8 m × 2 mm i.d. glass column packed with 3 per cent OV-17 of 80–100 mesh Chromosorb 750, was operated at an initial temperature of 100 °C for 1 min and was programmed to a final temperature of 225 °C at a rate of 3 °C min^{-1} and held at that temperature for the remainder of the analysis. The flow of helium carrier gas was set at 20 ml min^{-1} and the injection port temperature set at 200 °C. The glass jet separator and the ion source temperatures were set at 260 °C and 250 °C respectively. Data acquisition was under the control of the Finnigan 6110 data system. The mass range, 35–400 atomic mass units, was scanned at a rate of 2.1 s per scan and the mass spectra (*c.* 1000) stored on magnetic disc for subsequent analysis.

To test the effectiveness of their method of analysis for polyaromatic hydrocarbons in potable water Benoit *et al.*[11] prepared and analysed a control blank and carried out a recovery study of 32 selected polyaromatic hydrocarbons from XAD-2 resin. None of the compounds contained in the standard solution was detected in the concentrated extract from the control blank. This indicates that the XAD-2 resin is effective for the removal of these compounds from drinking water and that none of the reference compounds originates from the precleaned XAD-2 resin. However, when the amounts of polyaromatic hydrocarbons loaded and recovered are compared (Table 54) an average recovery of 84 per cent is observed with recoveries ranging from 57–100 per cent of the loaded material. The weighted average recovery was 88 per cent of loaded material. The rate of the unrecovered material was not established although, based on the results of the control blank, it is not likely that these materials were carried away by the effluent water.

Ottowa potable water samples were analysed in order to obtain some indication of whether the results are representative of the general background level of anthropogenic contamination. Aliquots of the reference standard solution (50 polyaromatic hydrocarbons and 5 oxygenated aromatic hydrocarbons) and the concentrated extracts from XAD-2 resin were analysed consecutively by gas chromatography–mass spectrometry under identical operating conditions. As is evident from Fig. 16(a) XAD-2 resin extract concentrate, as reconstructed from the total ion current, contained a multitude of poorly defined peaks. Complete mass spectra, free of extraneous ions, could rarely be obtained from such data despite the background subtraction routine possible with the

Table 54 The recovery of selected polycyclic hydrocarbons from Amberlite XAD-2 macroreticular resin.

Compound	Amount loaded (ng)	Fraction recovered
Naphthalene	625	0.57
2-Methylnaphthalene	1200	0.88
1-Methylnaphthalene	625	0.71
2-Ethylnaphthalene 2,6-Dimethylnaphthalene	>2300	>0.86
Biphenyl	775	0.66
1,3-Dimethylnaphthalene	975	0.81
2,3-Dimethylnaphthalene 1,4-Dimethylnaphthalene	>2250	>0.82
4-Phenyltoluene	600	0.85
Diphenylmethane 3-Phenyltoluene	>2075	>0.76
Acenaphthene	625	0.60
Bibenzyl	975	0.65
1,1-Diphenylethylene	1625	0.98
cis-Stilbene	575	1.0
2,3,5-Trimethylnaphthalene	700	0.75
3,3′-Dimethylbiphenyl	1625	1.0
Fluorene	750	0.89
4,4′-Dimethylbiphenyl	700	0.90
trans-Stilbene 9,10-Dihydrophenanthrene	>1075	>0.84
Phenanthrene Anthracene	>1700	>1.0
Triphenylmethane	800	0.87
Fluoranthene	650	1.0
Pyrene	725	1.0
1,2-Benzfluorene 2,3-Benzfluorene	>1550	>0.99
Triphenylene Benz(a)anthracene Chrysene	>1650	>1.0
Average	1110	0.84

From Benoit *et al.*[11] with permission.

data system and, hence, individual components of the concentrate were achieved from mass chromatograms (Fig. 16(b)) which were reconstructed from selected ion currents rather than the total ion current. Mass chromatograms for selected ions that were characteristic of the compound of interest were obtained by searching the accumulated data

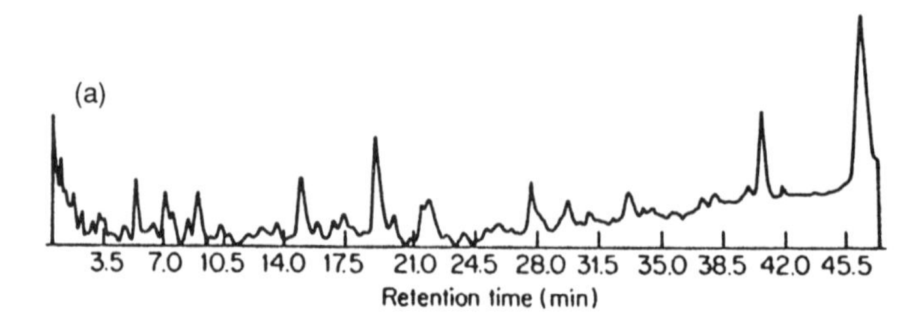

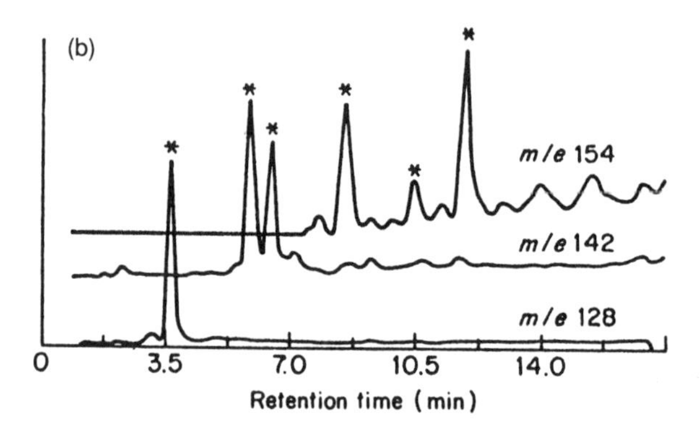

Fig. 16 Reconstructed gas chromatogram (total ion current) from Ottawa potable water, Amberlite XAD-2 resin extract. From Benoit *et al.*[11]

for the ion of interest and recording the abundance of this ion as a function of retention times. As an example, the mass chromatograms of three ions, *m/e* 128, *m/e* 142, and *m/e* 154, are superimposed in Fig. 16(b). The location of the peaks corresponding to the compounds of interest are indicated by asterisks in each chromatogram. For *m/e* 128 the asterisked peak corresponds to the molecular ion of naphthalene, for *m/e* 142 to the molecular ions of 2-methylnaphthalene and 1-methylnaphthalene, respectively, in order of increasing retention, and for *m/e* 154 to the molecular ions of biphenyl, 2-vinylnaphthalene, and acenaphthalene, respectively, in order of increasing retention times. The retention times for each standard were established by analysis of the reference standard solution and the data from the XAD-2 resin extracts were then searched for the ion of interest within the appropriate time region. In all instances, the molecular ion and the next most abundant ion were selected as the characteristic ions which are listed in Tables 55 and 56.

A compound was considered identified if the two characteristics of the compound of interest were found to elute from the column within the retention time window ($\pm$0.1 min) of the reference standard and

Table 55 Polycyclic aromatic hydrocarbons detected in Ottawa potable water sampled in January (2) and February (1) 1978.

Compound	Ions monitored		Rel. ret. time[a]	Concentration (ng litre^{-1})	
				1	2
Naphthalene	128	102	1.00	6.8	4.8
2-Methylnaphthalene	142	141	1.59	2.4	4.6
1-Methylnaphthalene	142	141	1.75	1.0	2.0
Azulene	128	102	1.90	n.d.	n.d.
2-Ethylnaphthalene	156	141	2.26	>0.70	2.1
2,6-Dimethylnaphthalene	156	141	2.32		
Biphenyl	154	153	2.30	0.70	1.1
1,3-Dimethylnaphthalene	156	141	2.51	1.9	1.1
2-Vinylnaphthalene	154	153	2.68	n.d.	n.d.
2,3-Dimethylnaphthalene	156	141	2.69	>0.68	14
1,4-Dimethylnaphthalene	156	141	2.69		
3-Phenyltoluene	168	167	2.74	0.20	1.5
Diphenylmethane	168	91	2.88	1.4	2.8
4-Phenyltoluene	168	167	2.94	0.20	3.7
Acenaphthylene	152	151	3.00	0.05	n.d.
Acenaphthene	154	153	3.25	0.20	1.8
Bibenzyl	182	91	3.41	1.9	1.5
1,1-Diphenylethylene	180	179	3.48	>7.4	n.d.
cis-Stilbene	180	179	3.59		
2,2-Diphenylpropane	196	181	3.62	n.d.	n.d.
2,3,5-Trimethylnaphthalene	170	155	3.71	0.65	5.2
3,3′-Dimethylbiphenyl	182	167	4.03	0.31	5.2
Fluorene	166	165	4.15	0.15	2.2
4,4′-Dimethylbiphenyl	182	167	4.18	0.57	7.0
4-Vinylbiphenyl	180	178	4.56	n.d.	n.d.
Diphenylacetylene	178	89	4.90	0.05	n.d.
9,10-Dihydroanthracene	180	179	5.03	0.66	n.d.
trans-Stilbene	180	179	5.20	>0.47	9.2
9,10-Dihydrophenanthrene	180	179	5.29		
10,11-Dihydro-5*H*-dibenzo(a,d)cycloheptane	194	179	6.03	0.40	n.d.
Phenanthrene	178	89	6.08	>0.52	2.2
Anthracene	178	89	6.14		
1-Phenylnaphthalene	204	203	6.77	n.d.	n.d.
1-Methylphenanthrene	192	191	7.06	n.d.	11
2-Methylanthracene	192	191	7.25	0.51	0.70
9-Methylanthracene	192	191	7.59	n.d.	0.70
9-Vinylanthracene	204	203	7.82	n.d.	n.d.
Triphenylmethane	244	167	8.04	n.d.	n.d.
Fluoranthene	202	101	8.41	0.55	1.9
Pyrene	202	101	8.90	0.53	1.7
9,10-Dimethylanthracene	206	191	8.99	0.19	n.d.

Table 55 cont.

Compound	Ions monitored		Rel. ret. time[a]	Concentration (ng litre^{-1})	
				1	2
Triphenylethylene	256	178	9.25	>0.08	n.d.
p-Terphenyl	230	115	9.44	n.d.	n.d.
1,2-Benzfluorene	216	108	9.64	n.d.	n.d.
2,3-Benzfluorene	216	108	9.73	n.d.	n.d.
Benzylbiphenyl	244	167	9.75	n.d.	n.d.
1,1′-Binaphthyl	254	126	10.95	n.d.	n.d.
Triphenylene	228	114	11.5		
Benz(a)anthracene	228	114	11.6	>8.1	3.3
Chrysene	228	114	11.8		

[a] Retention times are relative to the retention time of naphthalene (3.81 min).
From Benoit *et al.*[11] with permission.

Table 56 Oxygenated polycyclic aromatic hydrocarbons detected in Ottawa potable water sampled in January (2) and February (1) 1978.

Compound	Ions monitored		Rel. ret. time[a]	Concentration (ng litre^{-1})	
				1	2
Xanthene	182	181	4.83	0.20	0.10
9-Fluorenone	180	152	5.93	0.90	1.5
Perinaphthenone	180	152	7.70	0.28	0.15
Anthrone	194	165	7.90	1.4	n.d.
Anthraquinone	208	180	8.11	2.4	1.8
Naphthalene	128	102	1.00		

[a] Retention times are relative to the retention time of naphthalene (3.81 min).
From Benoit *et al.*[11] with permission.

to be in the relative abundance ratio (±20 per cent) observed in the mass spectrum of the pure compound. For most, but not all, compounds screened unique identification was possible. In some instances however, coeluting isomers yielding similar mass spectra could not be resolved sufficiently to allow unequivocal identification. Such coeluting isomers are grouped together in table 55 and are indicated by > beside the concentration values which is the sum of the contributions from all coeluting isomers. Furthermore it is emphasized that, because of the large number of compounds contained in the field sample extract, it was not possible to eliminate entirely from all the ion peaks of interest, contributions from possible interfering species. This was particularly true for methyl substituted polyaromatic hydrocarbons for which numerous positional isomers may elute within a narrow time window. In many cases only a small number of the possible positional isomers were available commercially and could be included in the reference standard. Hence, unequivocal identification of positional was often not possible.

Quantitative estimates of the detectable polyaromatic hydrocarbons and oxygenated polyaromatic hydrocarbons in Ottawa potable water were obtained by comparison of the areas of the two characteristic ion peaks (Tables 55 and 56), in the mass chromatograms of the reference standard and the field sample, respectively. The average of the concentrations for the two ions is presented in Table 55 (polyaromatic hydrocarbons) and Table 56 (oxygenated polyaromatic hydrocarbons) for the two water samples analysed. No corrections were made for incomplete recovery. Of the 50 polyaromatic hydrocarbons in the standard used by Benoit *et al.*[11] 38 are detected in at least one of the two potable water samples tested. In sample 1 (February 1978) 16 polyaromatic hydrocarbons ranging in concentration from 0.05 to 8.1 ng $litre^{-1}$ and in sample 2 (January 1978) 30 polyaromatic hydrocarbons (Table 55) ranging in concentration from 0.05 to 14 ng $litre^{-1}$ were detected. Twenty-eight polyaromatic hydrocarbons and four oxygenated polyaromatic hydrocarbons were detected in both samples analysed. The lower concentration of 0.05 ng $litre^{-1}$ represents the lower limit of detection of this method of analysis. There was an appreciable variation in the concentrations of most of the polyaromatic hydrocarbons detected in the two potable water samples; however, all compounds detected were found to be in the low ng $litre^{-1}$ range. This suggests that the observed concentrations of polyaromatic hydrocarbons and oxygenated polyaromatic hydrocarbons are representative of the background level of contamination. The mean concentration of polyaromatic hydrocarbons in samples 1 and 2 was 1.4 ng $litre^{-1}$ and 50.4 ng $litre^{-1}$.

Of the five oxygenated polyaromatic hydrocarbons in the standard, all five, ranging in concentration from 0.20 to 2.4 ng $litre^{-1}$ are detected in sample 1 and four, ranging in concentration from 0.1 to 1.8 ng $litre^{-1}$ are detected in sample 2 (Table 56). The mean concentrations and total weights of detected oxygenated polyaromatic hydrocarbons in the two samples were 1.0 and 5.2 ng $litre^{-1}$ for sample 1 and 0.91 and 3.7 ng $litre^{-1}$ for sample 2, respectively. It is noteworthy that for three of the oxygenated compounds (anthrone, anthraquinone, and 9-fluorenone) detected, the parent compound is also detected in the potable water sample. Thus, the oxygenated species could possibly originate from the oxidation of the parent compound in the aqueous media.

Junk and Richard[21] compared adsorption on XAD-2 resin and solvent extraction procedures for the preconcentration of polyaromatic hydrocarbons from groundwater and surface water samples. Compounds adsorbed on XAD-2 resin were desorbed with small volumes of ethyl acelate or benzene prior to gas chromatography.

Non-ionic detergents. Nickless and Jones[13] and Musty and Nickless[14,15] evaluated Amberlite XAD-4 resin as an extractant for down to 1 mg litre^{-1} of polyethylated or secondary alcohol ethoxylates (R)($OCH_2CH_2)_2OH$) surfactant and their degradation products from water samples. This resin was found to be an effective adsorbent for extraction of polyethoxylated compounds from water except for polyethylene glycols of molecular weight less than 300. Flow rates of 100 ml min^{-1} were possible using 5 g of resin, and interfering compounds can be removed by a rigorous purification procedure. Adsorption efficiences of 80–100 per cent at 10 μg litre^{-1} were possible for non-ionic detergents using distilled water solutions. The main purpose of the work of Nickless and Jones[13] was the investigation of secondary alcohol ethoxylate as it proceeds through the water system. Associated with this is polyethylene glycol, a likely biodegradation product. Alkylphenol ethoxylate was also considered but only as a possible interferent that should be differentiated in order to allow fuller characterization of secondary alcohol ethoxylate residues. Nickless and Jones[13] used in their study fairly pure secondary alcohol ethoxylate standards ($R(OCH_2CH_2)_nOH$) where n is 3, 5, 7, 9, and 12 also alkylphenol ethoxylate standards R-⟨◯⟩-($OCH_2CH_2)_nOH$ where n 2, 3, 9, and 22. The barium chloride phosphomolybdic acid spectrophotometric method[16,17] and also thin-layer chromatography[18] were used to determine polyoxyethylated compounds in column effluents. Nickless and Jones tested and efficiencies of three column materials for adsorption of secondary alcohol ethoxylate from solutions and found that Amberlite XAD-4 resin combined a high adsorption on the column with a high subsequent desorption of the ethoxylate from the column with ethano. Subsequently acetone was found to be a superior desorption solvent.

The XAD-4 results were very encouraging but the measured desorption was well over 100 per cent and was an indication of interfering compounds still present in the resin from its manufacture. These interfering compounds were polyanionic in nature and Nickless and Jones[13] devised a procedure involving successive washing with acetone–hexane (1:1), acetone, and ethanol of XAD-4 resin to remove impurities before its use in the analyses of samples. Results obtained by this procedure indicated a fairly rapid drop in adsorption efficiency of the resin for polyethylene glycols between 9EO and 3EO. It is not clear if the drop in efficiency is roughly linear with a shortening in chain length, but it appears that the resin might be limited to the study of polyethylene glycols in water for chain length greater than 7EO.

The ability of XAD-4 resin to adsorb with high efficiency polyethoxylated compounds at very low concentrations is very important, if it is to be used for the examination of all types of water systems. When the concentration is very low then this necessitates the processing of large volumes of water in order to obtain sufficient material for characterization using methods such as infra-red, ultraviolet, and NMR spectroscopy, as well as liquid chromatography. The performance of the resin in this respect was evaluated by using solutions of a mixture of three model compounds, secondary alcohol ethoxylate 9EO, alkylphenol ethoxylate 9EO, and polyethylene glycol 9EO. Recoveries varied from 82 per cent at the 0.01 mg $litre^{-1}$ level to 98 per cent at the 1 mg $litre^{-1}$ level.

Nickless and Jones[13] continued their study of Amberlite XAD-4 resins by examining polyethoxylated materials before and after passage through a sewage works. Samples from the inlet and outlet of the sewage works and from the adjacent river were subjected to a three stage isolation procedure and the final extracts were separated into a non-ionic detergent and a polyethylene glycol. The non-ionic detergent concentration was 100 times lower in the river than in the sewage effluent. Thin-layer chromatography and ultraviolet, infra-red and NMR spectroscopy were used to identify, in the non-ionic detergent component, alkylphenol ethoxylates (the most persistent), secondary alcohol ethoxylates, and primary alcohol ethoxylate.

The three stage isolation procedure was carried out as follows.

Stage 1. The water sample was first passed down a column of XAD-4 resin to extract surface active materials on to the resin. The organics adsorbed from the water sample by the XAD-4 resin were eluted off with four solvent systems to give two fractions:

(1) 2 ml of methanol–water (1:1) followed by 20 ml of methanol;

(2) 50 ml acetone, followed by 20 ml of acetone–*n*-hexane (1:1).

Each fraction was collected and evaporated to dryness in a stream of filtered air on a hot water bath. Fraction 2 contained most of the polyethoxylated material but significant amounts were also present in fraction 1. The residue from fraction 1 was treated with 10 ml of acetone, decanted into a small tube, centrifuged, and the clear acetone layer poured into the beaker containing fraction 2. The combined extracts were again evaporated to dryness.

Stage 2 — liquid-solid chromatography. The polyethoxylated material contained in fraction 2 from the XAD-4 elution will still be a relatively minor part of the residue. The indications were that most of the unwanted organic compounds extracted form the river water would be medium to non-polar in character. There would therefore be overlap with the medium polar oligomers of secondary alcohol ethoxylate and alkylphenol ethoxylate (i.e. those with only 1, 2 or 3 ethylene oxide units per molecule). Silica gel was found to be the best chromatographic adsorbent and the correct choice of eluting solvent strength is important for efficient separation, a (7:3) ethyl acetate; benzene extract of fraction 2 was poured down a column of silica gel. The ethyl acetate–benzene insoluble residue contained highly polar polyethoxylated material and was dissolved and loaded on to the column when more polar solvents were used later in the eluting procedure. Most of the unwanted organic compounds came through in fractions 1 and 2 and were discarded. Fractions 3 and 4 (less than 1 per cent of total) were combined and kept for thin layer chromatographic examination. Fractions 5 and 6 were combined and usually contained greater than 98 per cent of the total polyethoxylated material present in the original extract, together with some highly polar compounds imparting a faint yellow colour to the solution.

Stage 3. The combined fractions 5 and 6 obtained from liquid solid chromatography needed further separation for two reasons:

(1) the fraction still contained significant amounts of other organic compounds which were found to be mainly acidic in character, and
(2) meaningful results could only be obtained from spectroscopic examination if the polyethoxylated material is divided into the non-ionic detergent components and the polyethylene glycol component.

Jones and Nickless[19] solved both of these problems in one operation using liquid chromatography based on a procedure by Nadeau and Waszeciak.[20] With this procedure using water as a stationary phase (on Celite) and chloroform–benzene as the mobile phase, very good separations of polyethylene glycol from secondary alcohol ethoxylate and alkylphenol ethoxylate can be obtained. By simply changing the stationary phase to dilute sodium hydroxide solution, unwanted acidic compounds can be removed with the polyethylene glycol fraction.

Having separated the mixture into a secondary alcohol ethoxylate plus alkylphenol ethoxylate and polyethylene glycol components, Jones and Nickless examined these fractions using thin-layer chromatography[18] and ultraviolet infra-red, and NMR spectroscopy.

In the thin layer chromatographic separations ethyl acetate–acetic acid–water, 4:3:3, was used for quantitative information, where a compact spot is obtained for non-ionic detergents (secondary alcohol ethoxylate and alkylphenol ethoxylate combined) and an elongated spot of lower R_f value for polyethylene glycol. Ethyl acetate–acetic acid–

water, 70:15:15, was used for information on the molecular weight distribution of secondary alcohol ethoxylate and alkylphenol ethoxylate. Alkylphenol ethoxylate gives a 'string' of well resolved spots and secondary alcohol ethoxylate a long unresolved streak. These patterns will be superimposed if both are present in the residue.

Ultraviolet spectroscopy was used to evaluate alkylphenol ethoxylates as only this types of compound gives a peak at 277 nm with a characteristic shoulder at 285 nm. Chloroform was the solvent used and after measurement the sample can be reconcentrated in a stream of air back to 0.5 ml. The infra-red spectra of secondary alcohol ethoxylates and alkylphenol ethoxylates isolated from samples are very similar with the broad strong peak at 1100–1120 cm^{-1} characteristic of a polyethoxylate grouping. The only clearly recognizable difference between them is the very sharp aromatic peak present in the alkylphenol ethoxylate spectrum at 1500–1505 cm^{-1}. NMR spectroscopy was able to distinguish between secondary alcohol ethoxide and primary alcohol ethoxylate.

Organophosphorus insecticides. Paschal *et al.*[22] hve discussed the preconcentration and determination of parathion-ethyl and parathion-methyl in run-off water using high performance liquid chromatography. The organic compounds are concentrated on and XAD-2 resin before analysis by reverse phase, high performance liquid chromatography. Detection limits were found to be approximately 2–3 mg $litre^{-1}$. These workers examined the possible interference in the method from other agricultural chemicals and organic compounds commonly occurring in water. This method is based on the use of Rohm and Haas XAD-2 macroreticular resins. Organics in water can be sorbed on a small column of resin, and the sorbed organics then eluted by diethyl ether. After evaporation of the eluate, the concentrated organics can be determined by chromatography. In addition to the obvious benefit of 100- to 100-fold concentration, this method offers the possibility for on-site sampling.

Apparatus:

Gas chromatograph, six port injector, a Whatman prepacked microparticle reverse phase column (Partisil ODS) and a variable wave-length detector.

Reagents and materials:

Macroreticular resin XAD-2, purified by Soxhlet extraction as described by Junk[23] and stored under pure methanol.

Pesticide grade acetonitrile as received.

Diethyl ether, glass distilled before use.

Distilled water, glass distilled before use.

Preparation of standards:

Make up microlitre amounts of 100 mg litre^{-1} stock solutions of organophosphorus insecticides in methanol and dilute with distilled water to 100 ml. Pass the diluted standards through a 10 cm column of purified XAD-2 resin, at a rate of 4–6 ml min^{-1}. After the last of the dilute aqueous standards had passed through the column, remove most of the water clinging to the resin by gentle vacuum aspiration. Pass 30 ml of glass distilled diethyl ether through the column at 2–3 ml min^{-1} then remove the last of the ether by passing dried purified nitrogen through the column. Dry the ether by shaking with 2 g of anhydrous sodium sulphate and evaporate to dryness using a rotary evaporator at temperatures not exceeding 35 °C. Dissolve the residue in 100 ml of pesticide grade acetonitrile, and chromatograph the resulting solution on a Partisil ODS reverse phase column at 2.40 ml min^{-1} with 50 per cent acetonitrile/water mobile phase. An average recovery of 99 per cent was obtained through the procedure.

Procedure:

Prepare extracts of 2 litre samples as in the above procedure and evaluate chromatograms to establish calibration curves for the parathions.

To evaluate the efficiency of extraction of XAD-2 resin for trace organics in run-off water, Paschal *et al.*[22] ran a sample of water through the procedure and obtained a large number of peaks in the 2–3 min region. A number of relatively polar compounds elute early in the chromatograph, with relatively few peaks in the 3–10 min region of the chromatogram. On changing from 50 per cent acetonitrile to 100 per cent acetonitrile to regenerate the column, several more peaks were eluted, apparently consisting of less polar materials strongly adsorbed under the conditions of the procedure. No interference was obtained from these strongly adsorbed compounds, although it was found to be useful to regenerate the column with 100 per cent acetonitrile after every five to six runs to ensure reproducibility.

The retention times for parathion-methyl and parathion-ethyl, obtained from volumetric dilutions of methanolic standards with acetonitrile, were 3.45 and 4 min, indicating no interference from naturally occurring organics in the run-off water. Spiked samples of run-off water were prepared containing parathion-ethyl and parathion-methyl. A typical chromatogram for such a spiked sample is shown in Fig. 17. The parathions are well separated, with no observed interference from organics already present in the water. Calibration curves were prepared from a set of standards containing 10–120 μg litre^{-1} parathion in methanol. Atrazine was added as an internal standard to the concentrated extract. Ratio of speak heights or areas of parathions

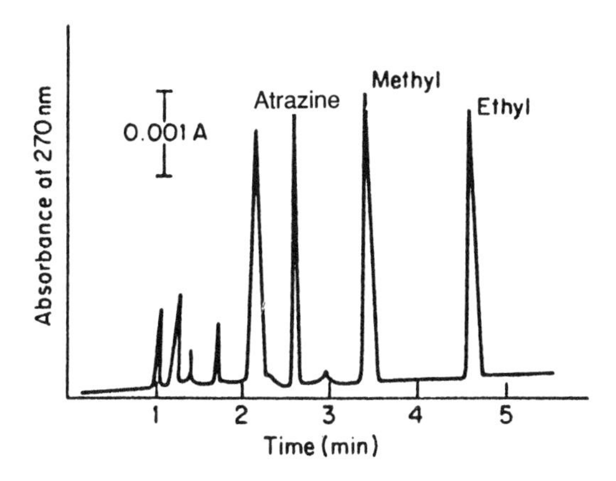

Fig. 17 Chromatogram of spiked run off water extract. From Paschal *et al.*[22]

to those of atrazine were plotted versus concentration. Good linearity was obtained over the range of concentration examined for both parathions. In order to evaluate the accuracy and reproducibility of the method, a series of solutions was prepared in run-off water with concentrations of parathions in the range of the calibration curve. The results of this study are given in Table 57. The lower limit of detection was calculated from those data to be 3.1 and 2.9 ng for parathion-methyl and parathion-ethyl, respectively.

Table 58 shows the effect of potential interference by other agricultural chemicals and organics commonly occurring in natural water. Wavelengths chosen for measurement were at or near the absorbance

Table 57 Reproducibility of method.

Taken (μg litre^{-1})	Found (μg litre^{-1})[a]	s.d.	Relative s.d. (per cent)
Parathion-methyl			
15.0	14.8	0.45	3.0
37.5	37.1	1.07	2.8
75.0	75.9	0.73	1.0
112.5	112.7	2.56	2.3
Parathion-ethyl			
10.0	9.9	0.37	3.7
25.0	24.6	1.40	5.6
50.0	49.3	0.97	1.9
75.0	75.0	2.40	3.2

[a] Average of six determinations.
From Paschal *et al.*[22] with permission.

Table 58 Interference study.

Compound	Relative retention (parathion-methyl = 1.00)	Wavelength measured (nm)
Aroclor 1260	3.94–5.88 multiple peaks	225
Atrazine	0.75	265
Azinphos-ethyl	1.14	285
Alachlor	0.89	235
Carbaryl (Sevin)	0.69	280
Carbofuran	0.61	270
Chloramben	0.26	240
Chlorpyrifos	2.01	290
p,p'-DDT	2.78	235
DEHP	1.59	235
Dialifor	1.61	290
Diazinon	1.30	245
Dyfonate (fonofos)	1.36	240
Fenitrothion	1.18	265
Methoxychlor	1.72	225
p-Nitrophenol	0.72	310
Phosmet	0.93	230
Phorate	1.30	220
Propachlor	0.67	260
2,3,5-T	0.28	250
Trifluralin	0.58	270

From Paschal *et al.*[22] with permission.

maxima for the compounds as determined by UV scans from 350 to 200 nm. If a potentially interfering compound showed a retention time near one of the parathions, then chromatography was performed with detection at 270 nm. Of the compounds investigated only fonofos (ethyl-*S*-phenylethyl phosphonothiolothionate) interferes at 270 nm. However, if the wavelength of detection is changed to 280 nm, the interference is overcome.

Various workers[24-30] have reported on the use of XAD-2 for the extraction of fenitrothion from water. Mallet *et al.*[27] used an automated gas chromatographic system which consisted of a gas chromatograph mounted with an automatic sampler interfaced to an integrator. A Melpar flame photometric detector (phosphorus mode) was connected with the flame gas inlets in the reverse configuration to prevent solvent flame-out. The detector was maintained at 185 °C and flame gases were optimized with flow rates (ml min^{-1}) as follows: hydrogen, 80;

oxygens, 10; air, 20. A 1.8 m × 4.0 mm i.d. U shaped glass column packed with 4 per cent (w/w) OV-101 and 6 per cent (w/w) OV-210 on Chromosorb W AW DMCS, 80–100 mesh, was used. Nitrogen was used as carrier gas at a flow rate of 70 ml min^{-1}. A column temperature of 195 °C sufficiently resolved fenitro-oxon from its parent compounds. The injection port temperature was set at 225 °C. The water sample was passed through an XAD-2 column, which was subsequently eluted with ethyl acetate. The ethyl acetate extract was examined by gas chromatography.

In an alternative method involving thin layer chromatography, and fluorimetric analysis, a 1 liter water sample was extracted with two 50 ml portions of chloroform, which were collectively dried through an anhydrous sodium sulphate (50 g) column. The chloroform was replaced with ethyl acetate on a flash evaporator, carefully reduced to 4–5 ml and made up to the mark with ethyl acetate in a 10 ml volumetric flask. Aminofenitrothion was recovered by this method from environmental water using an XAD-2 column. Recoveries were in the range 87–118 per cent at an average flow rate of 153 ml min^{-1}. The relative error, *c.* 10 per cent, is normal at a concentration of 50 μg $litre^{-1}$ when using in-situ fluorimetry. Under similar conditions, fenitro-oxon can also be recovered with good yields. The procedure was adapted to the simultaneous analysis of the parent compound and its two derivatives by gas liquid chromatography. With XAD-2 resin, conditions such as flow rate and column length are crucial to obtain good recoveries. If a 10 × 1.9 cm i.d. column is used the maximum flow rate is limited to *c.* 50 ml min^{-1}, which is easily sustained by gravity flow.

Various good recoveries are obtained when using XAD-2 for extracting spiked lake water with fenitro-oxon. Relative standard deviations of 5.1–6.4 per cent are very good. The overall average recovery (99.7 per cent) of the three compounds by the conventional serial solvent extraction procedure is somewhat better than by the XAD method (95.3 per cent). Reproducibilities are all better with the exception of fenitro-oxon. Volpe and Mallet[30] developed a method for preconcentrating and determining down to 0.5 ng of fenitrothion and five fenitrothion derivatives in water by adsorption on XAD-4 and XAD-7 resins, followed by solvent elution and gas liquid chromatography of the extract.

Puijker *et al.*[31] preconcentrated and separated organic compounds containing phosphorus and sulphur from water on to XAD resin, then reduced the compounds with hydrogen at 1100 °C. The resulting phosphine and hydrogen sulphide were separated on chromatographic column, and detected at 526 and 384 nm respectively, in a flame photometer. The detection limits for phosphorus and sulphur were 0.1 ng and 1 ng respectively. Frobe *et al.*[32] preconcentrated organophosphorus insecticides on Amberlite XAD-4 prior to determination of phosphorus by oxygen flask combustion.

Organochlorine pesticides and herbicides. XAD-2 resin has been used to preconcentrate a range of organochlorine pesticides including dieldrin[33,34] prior to desorption with hot solvents (e.g. acrylonitrile[34]) and gas chromatography. Levesque and Mallet[35] used preconcentration on XAD-4 and XAD-7 resins followed by solvent extraction of adsorbate and gas chromatography using a nitrogen phosphorus detector to determine aminocarb herbicide and some of its degradation products in water. Similarly, aldicarb and its oxidation products, aldicarb sulphoxide and aldicarb sulphone, have been determined using XAD-2 resin and high performance liquid chromatographic finish.[36] Harris *et al.*[37] preconcentrated kepone at low concentrations using XAD-2 resin. Sundaram *et al.*[38] preconcentrated carbamate herbicides on XAD-2 resin and desorbed them with a small volume of ethyl acetate or acetone prior to gas chromatography using an NP specific detector. Chloromethoxynil, bifenox, and butachlor have been preconcentrated by similar techniques which are capable of achieving detection limits of 0.2 μg litre^{-1}.[39]

Rees and Au[40] used small columns of XAD-2 resin for the recovery of ambient trace levels of pesticides (organochlorine also organophosphorus pesticides, triazine and chlorophenoxy acid herbicides, phthalate esters, and polychlorinated biphenyls) from water samples at concentrations ranging from 0.001 to 50 μg litre^{-1}.

Reagents:

Solvents distilled in glass, glass wool and sodium sulphate, solvent washed prior to use.

Diethyl ether, redistilled daily over metallic sodium to remove alcohol preservative.

XAD-2 resin, 20–60 mesh, purify by sequential extraction with methanol, acetonitrile, and diethyl ether in a Soxhlet extractor for 8 h per solvent. Store purified resin under methanol.

Florisil, PR grade, 60–80 mesh, store at 130 °C.

Standard solutions made up in hexane or benzene. Appropriate dilutions for fortifications made in acetone.

Apparatus:

Resin column, Pyrex's 1.0 × 20 cm with Teflon stopcock and 1 litre integrator reservoir.

Concentration flask, 250 ml r.b. flask with 10 ml graduated conical extension.

Florisil column, Pyrex, 0.6 × 20 cm with Teflon stopcock and 100 ml reservoir.

Gas chromatographic conditions:

Organochlorine pesticides: Pyrex column, 190 cm × 2 mm, packed with 11 per cent OV 17/QFI on Gas Chrom Q was operated at 215 °C, Ni 63 detector was operated at 300 °C.

Chlorophenoxy Acids: The same conditions as above except for the column temperature (170 °C) while for PCBs the column packing was 3 per cent Dexsil 300 on Chromosorb W.

Phthalates: Flame-ionization detector at 300 °C, column temperature programmed from 90 °C to 250 °C at 10 °C min^{-1}. Packing Carbowas 20 M on Chromosorb W.

Organophosphates: Flame photometric detector in the reversed flow mode (air: 40 ml min^{-1}; oxygen, 20 ml min^{-1}; hydrogen: 2.00 ml min^{-1}) operated at 200 °C. Column (3 per cent Dexsil 410) programmed from 100 to 220 °C at 7 °C min^{-1}.

Triazines: Column packing 6 per cent Carbowax 20 M on Chromosorb W, conditioned at 260 °C. Oven temperature, programmed from 190 to 250 °C at 4 °C min^{-1}. AFID detector, operated at 260 °C.

Procedure:

Set up the resin column and plug the tower end with glass wool. Add XAD-2 resin as a methanol slurry until a 6 cm bed is formed. Insert a second glass wool plug to cap off the bed. Drain the methanol until the level reaches the top of the bed; then rinse the bed with 3 × 30 ml portions of pre-purified redistilled water in order to wash off the methanol and aid in reduction of air bubble formation. For higher concentrations of pesticides, add 1 litre of pre-purified distilled water to the reservoir, fortified with the appropriate acetone solution (1 ml) swirl to mix for 1 min, then drain through the column at approximately 35–40 ml min^{-1}. Further rinse the reservoir with 30 ml pre-purified distilled water, which is also drained through the column. For lower concentrations, where 20 litres of water are used, carry out fortifying in one-gallon pre-rinsed bottles as alternating supplies. Transfer fortified water to the reservoir by syphoning through 1/8″ Teflon tubing.

After all the water has passed through the resin, allow the bed to drain for 5 min, then run absolute diethyl ether through the resin and collect in a 250 ml separatory funnel, until no more water in coeluted. Then close the column tap and allow the dry ether (approx. 20 ml) to stand in the resin bed for approximately 10 min. Then run off the ether and add to the contents of the separatory funnel. Repeat the equilibration with two further 20 ml portions of ether, adding each eluate to the separatory funnel.

After separation, run off the water in the funnel and discard. Dry the remaining ether by passing through sodium sulphate and collect in a 250 ml concentrating flask. Vacuum rotary evaporate the extract (30 °C) to 0.5 ml, blow gently to dryness, make up in hexane, and submit for gas chromatographic analysis under the conditions mentioned above. In the case of organochlorine pesticides and PCBs clean up extracts using a 16 cm hexane slurry packed Florisil column prewashed with 40 ml hexane. Transfer the sample to the column then elute with 15 ml 25:75 (v/v) dichloromethane: hexane, containing PCBs, lindane, heptachlor, aldrin, DDT group, and 15 ml dichloromethane containing heptachloroepoxide, dieldrin, endrin. For chlorophenoxy acid recoveries, acidify fortified solutions to pH 2, before resin extraction and methylate the concentrated eluates with diazomethane/diethyl ether prior to Florisil clean up and gas chromatography.

The adsorption efficiency of the resin was tested by running large samples (47 litres) of natural river water through two successive cartridges. Individual desorption of each cartridge showed that there was no trace of carry through and the organochlorines and PCBs were completely adsorbed on the first cartridge, at levels of 3 ng litre^{-1} for PCBs and 9 ng litre^{-1} of pp-DDD. Thus concentrations of 0.1 ng litre^{-1} of PCB and 0.1 ng litre^{-1} of organochlorinated pesticides can be detected in natural waters.

Polychlorinated biphenyls. XAD-2 and XAD-4 macroreticular resins have been used to preconcentrate polychlorobiphenyls prior to solvent desorption and analysis in amounts down to 0.4 ng litre^{-1} using either electron capture gas chromatography[42-51] or high performance liquid chromatography.[41] Musty and Nickless[147] used XAD-4 resin to preconcentrate polychlorobiphenyls from 1-litre samples of natural water prior to desorption with diethyl ether: hexane (1:1) and gas chromatography.

Miscellaneous organic compounds. Various other applications of non-polar types of XAD resins to the preconcentration of organics in water are reviewed in Table 59.

Sea-water – general discussion Macroreticular resins have also been used for the collection of trace organics in sea-water. An excellent early review of the properties of the various XAD resins, along with comparisons with EXP-500 and activated carbon, can be found in Gustafson and Paleos.[91] Riley and Taylor[92] have studied the uptake of about 30 organics from sea-water on to XAD-1 resin at pH 2–9. At the 2–5 μg litre^{-1} level none of the carbohydrates, amino acids, proteins or phenols investigated were adsorbed in any detectable amounts. Various carboxylic acids, surfactants, insecticides, dyestuffs, and especially humic acids are adsorbed. The humic acids retained on the XAD-1 resin were fractionated by elution with water at pH 7, 1 M aqueous ammonia, and 0.2 M potassium hydroxide.

Osterroht[93,94] studied the retention of non-polar organics from sea-water on to macroreticular resins. Each of the XAD resins has slightly different properties and should collect a slightly different organic fraction from sea-water. The major differences between the resins is in the degree of their polarity. The macroreticular resins should be useful in the analysis of particular classes of compounds; the stumbling block will be the determination of efficiencies. Earlier experience, admittedly with early versions of the resins, was that the collection of organic materials from water was far from complete. Once the properties of the resins are well understood, the analysis of at least some classes of compounds may quickly become routine.

Applications

Polychlorinated biphenyls and chlorinated insecticides. Musty and Nickless[95] used Amberlite XAD-4 for the extraction and recovery of chlorinated insecticides and PCBs from sea-water. In this method a glass column (20 × 1 cm) was packed with 2 g XAD-4 (60–85 mesh), and 1 litre of water (containing 1 part per 10^9 of insecticides) was passed through the column at 8 ml min^{-1}. The column was dried by drawing a stream of air through, then the insecticides were eluted with 100 ml ethyl ether–hexane (1:9). The eluate was concentrated to 5 ml and was subjected to gas chromatography on a glass column (5.5 ft (1.7 m) × 4 mm) packed with 1.5 per cent OV-17 and 1.95 per cent QG-1 on Gas-Chrom Q (100–120 mesh). The column was operated at 200 °C, with argon (10 ml min^{-1}) as carrier gas and a ^{63}Ni electron capture detector (pulse mode). Recoveries of BHC isomers were 106–114 per cent; of aldrin, 61 per cent; of DDT isomers, 102–144 per cent; and of polychlorinated biphenuls 76 per cent.

Elder[96] determined PCBs in Mediterranean coastal waters by adsorption on to XAD-2 resin followed by electron capture gas chromatography. The overall average PCB concentration was 13 ng $litre^{-1}$.

Amberlite XAD-2 resin is a suitable adsorbent for polychlorinated biphenyl and chlorinated insecticides (DDT and metabolites, dieldrin) in sea-water. These compounds can be suitably eluted from the resin prior to gas chromatography.[97] Picer and Picer[98] evaluated the application of XAD-2; XAD-4, and Tenax macroreticular resins for concentrations of chlorinated insecticides and polychlorinated biphenyls in sea-water prior to analysis by electron capture gas chromatography. The solvents used eluted not only the chlorinated hydrocarbons of interest but also other electron capture sensitive materials, so that eluates had to be purified. The eluates from the Tenax column were combined and the non-polar phase was separated from the polar phase in a glass

Table 59 Application of non-polar XAD resins (XAD-2, XAD-4), to preconcentration of organics from waters

Substance	Resin	Method of desorption	Analytical finish	Type of water sample	Detection limit	Ref.
Total organic halogen	XAD	Solvent elution	Schoniger combustion–potentiometric titration of halogen	Natural	1–2 μmol litre^{-1}	52
Chloroethanes	XAD-4	–	–	Natural	–	53
Humic substances	XAD-2	–	–	Natural	–	54
Haloforms	XAD-2, XAD-4	Solvent elution	GLC	Natural	–	55
Haloforms	Acetylated XAD-2	Pyridine	GLC	Natural	–	56,57
Haloforms	XAD-4	Ethanol	GLC	Natural	–	58
Alkylphosphates, alkylthiophosphates	XAD-4	Solvent	GLC	Natural	–	59
Phenols	XAD-2	Pyridine	Silation GLC	Natural	0.2 mg litre^{-1}	60
Phenols	XAD-2	Pyridine	–	Natural	–	61
Phenols	XAD-4 (aminated with trimethylamine)	Diethyl ether	Methylation–GLC	Natural	–	62
Phenol, *p*-chloropenol	XAD-4	–	–	Natural	–	63
Nitrilo acetic acid	Dowex 1	Formic acid	Propylation–GLC	Natural	0.01 μg litre^{-1}	64
Chlorobenzenes	XAD-2	Solvent	GLC	Natural	1-10 μg litre^{-1}	65

Miscellaneous hydrocarbons phenols, alkyl phthalates, polyaromatic hydrocarbons	XAD-2	Solvent elution	GLC	Natural	–	66
Miscellaneous carboxylic acids, alcohols, amines, sucrose, amino acids, quinoldic acid	XAD-2	–	–	Fresh natural	–	67–70
Miscellaneous, alcohols, esters, aldehydes, alkylbenzenes, polynuclear hydrocarbons, chlorocompounds	XAD-4	–	–	Waste and potable	–	71
Hydrocarbons	XAD-2	–	–	Natural	–	72
Miscellaneous alkanes, alkylbenzenes, haloforms, polynuclear hydrocarbons, ketones, alcohols, phenols, carboxylic acids	XAD-2	–	–	Potable	–	73
Miscellaneous alkanes, alkylbenzenes, chlorocompounds, alcohols, ketones, esters, bromocompounds	XAD-4	–	–	Potable waste	–	74

Table 59 cont.

Substance	Resin	Method of desorption	Analytical finish	Type of water sample	Detection limit	Ref.
Miscellaneous alkylbenzenes, alkanes, phenols, carboxylic acids chlorocompounds	XAD-2	–	–	Natural	–	75
Miscellaneous phenols, carboxylic acids, aldehydes, alcohols	XAD-2	–	–	Waste	–	76
Miscellaneous	Misc. XAD	Solvent extraction	–	–	–	77
Miscellaneous	XAD-2	Solvent extraction	–	Potable	–	78
Miscellaneous	Misc. XAD	Solvent extraction	–	Trade effluents	–	79
Miscellaneous	Misc. XAD	–	–	Natural	–	80
Miscellaneous	Misc. XAD	Thermal desorption	GLC	Potable	–	81

Miscellaneous	XAD-2	–	GLC, GLC–MS	Pulp effluent	–	82
Miscellaneous	XAD-2	–	–	River	–	83
Miscellaneous	XAD-4	Solvent extraction (CS_2)	GLC–MS	Surface waters	0.1 μg litre^{-1}	84
Miscellaneous	XAD-2	–	–	Well water	mg litre^{-1}	85
Miscellaneous	XAD-2 XAD-4	CS_2 extraction	GLC–MS	Surface water	–	86
Miscellaneous	XAD (Misc)	Solvent extraction	–	Natural	–	87
Miscellaneous	XAD-1	–	–	Natural	–	88
Miscellaneous	XAD-2	–	–	Natural	–	89
Miscellaneous acidic, humic and neutral types	XAD-2	–	–	Natural	–	90

separating funnel. Then the polar phase was extracted twice with *n*-pentane. The *n*-pentane extract was dried over anhydrous sodium sulphate, concentrated to 1 ml and cleaned on an alumina column using a modification of the method described by Holden and Marsden.[99] The eluates were placed on a silica gel column for the separation of PCBs from DDT, its metabolities, and dieldrin using a procedure described by Snyder and Reinert[100] and Picer and Abel.[101]

Picer and Picer[98] investigated the preconcentration from 10-litre samples of sea-water of 0.1–1.0 μg litre^{-1} chlorinated pesticides (DDT, DDE, TDE, and Dieldrin), and 1–2 μg litre^{-1} PCB (Aroclor 1254). Interestingly, for the elution of all chlorinated hydrocarbons 25 ml polar solvent were required, and 50 ml *n*-pentane was used as the re-extractant. Mirex was used as an internal standard, added to the eluate after the percolation of the polar solvent through the resin column. Hence this internal standard shows only the loss of chlorinated hydrocarbons during the re-extraction, alumina clean-up, and silica gel separation. The recovery of Mirex during these steps varied between 80 per cent and 90 per cent. Losses of the investigated chlorinated hvdrocarbons during these steps were 10–30 per cent for about 10 ng pesticides. The experimental set-up is obviously capable of determining hydrocarbons in a 10 litre sea-water sample at levels far below 1.0 ng litre^{-1} for pesticides and 10 ng litre^{-1} for PCBs.

Of the three macroreticular resins investigated XAD-2 was the best using methanol as elution solvent. Very unsatisfactory solvent blanks were obtained using Tenax resin. These workers conclude that application of macroreticular resins for the adsorption of chlorinated hydrocarbons from water samples and their determination after elution with different solvents has revealed several limitations. When water samples were spiked at levels close to the reported concentrations in sea-water, the recovery of the investigated chlorinated hydrocarbons was low and unpredictable.

Organochlorine insecticides have been preconcentrated on XAD-2 resin prior to gas chromatographic analysis.[102–104] Gomez-Belinchon *et al.*[105] carried out an intercomparison study of three methods involving adsorption on XAD-resin, liquid–liquid extraction (see Chapter 7), and adsorption on polyurethane foam (see Chapter 4) for the preconcentration of polychlorobiphenyls, hydrocarbons, and fatty acids from sea-water. Sample sizes of 300–400 litres were used enabling very high preconcentration factors to be achieved. All three methods gave similar quantitative results, but there were qualitative differences which suggested that in this instance at least liquid–liquid extraction should be the method of choice.

Organophosphorus insecticides. The degradation products of the organophosphorus insecticides can be concentrated from large volumes of water by collection on Amberlite XAD-4 resin for subsequent analysis.[106] These are certainly not the only references to methods for the phosphorus-containing insecticides in natural waters; however, most of the work has been done in fresh water and at concentrations very much higher than those to be expected in sea-water.

Chlorinated aliphatic compounds. Dawson *et al.*[107] have described samplers for large volume collection of sea-water samples for chlorinated hydrocarbon analyses. The samplers use the macroreticular absorbent Amberlite XAD-2

Azarenes. Shinohara *et al.*[108] have described a procedure based on gas chromatography for the determination of traces of two, three, and five ring azarenes in sea-water. The procedure is based on the concentration of the compounds on Amberlite XAD-2 resin, separation by solvent partition,[109] and determination by gas chromatography–mass spectrometry with a selective ion monitor. Detection limits by the flame thermionic detector were 0.5–3.0 ng and those by gas chromatography–mass spectrometry were in the range 0.02–0.5 ng. The preferred solvent for elution from the resin was dichloromethane and the recoveries were mainly in the range 89–94 per cent.

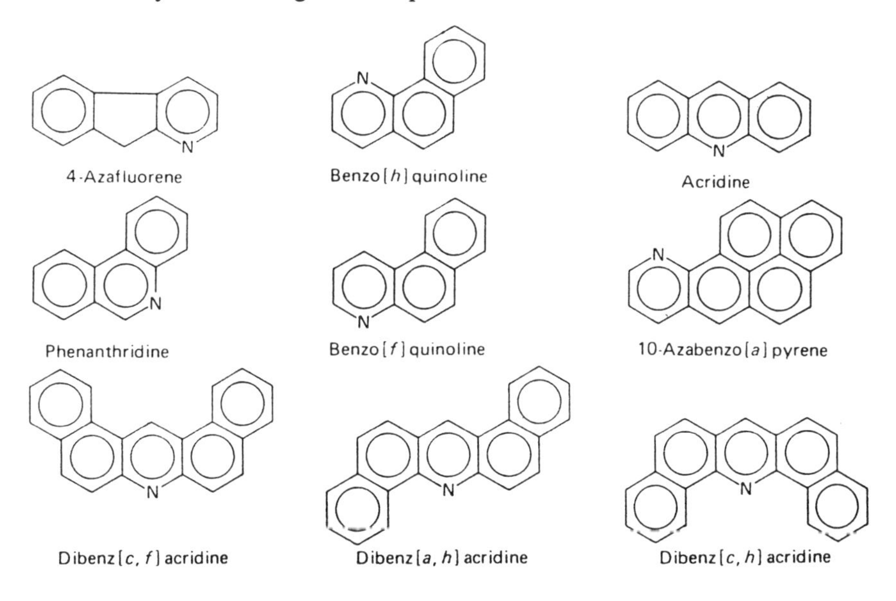

Humic and fulvic acids. Several workers have used macroreticular resins, usually XAD-2, to collect high molecular weight humics from sea-water.[110–112]

Sterols. The sterols differ from the other compounds in that no class reaction has been proposed for the measurement of total sterols. Instead, various fractionation methods, usually derived from the biochemical literature, have been adapted to the concentrated materials collected from sea-water. Certain of the more important sterols, particularly those used in the evaluation of water quality, have been determined by the use of a compound-specific reaction, after concentration from solution. Thus Wun *et al.*[113,114] measured coprostanol, a faecal sterol in sea-water, after collection and separation, by extraction using liquid–liquid partitioning or extraction on a column of Amberlite XAD-2 resin.

Organosulphur compounds. Adsorption on XAD-2 and XAD-4 resins followed by solvent desorption and head space gas chromatography has been employed for the preconcentration and determination of volatile organosulphur compounds in estuary and sea waters.[115]

3.1.2 Tenax GC

Non-saline waters The Tenax materials manufactured by Akzo Research Laboratories are a range of porous polymers based on various polymers. For example, Tenax GC is based on 2,6-diphenyl-*p*-phenylene oxide. After adsorption of the organics from a large volume of water on to the resin, desorption can be achieved either by solvent extraction or thermal elution. Organic porous polymers have recently become popular in concentrating air and water pollutants. Among these porous polymers Tenax GC has become widely accepted in air and water analysis. The sorption power of Tenax is nearly unaffected by water and repeated re-use of the sorbent is possible. Also, it has high thermostability (thermal elution) and its compatability with alcohols, amines, amides, acids, and bases with good recovery characteristics make Tenax very suitable as sorbent medium in water analysis. A disadvantage, however, is small break through volumes for light hydrocarbons and low molecular weight polar compounds thus limiting the sample volumes. A solution to this problem is to use another more active sorbent in series with Tenax for collecting low molecular weight compounds breaking through on Tenax.

Applications

Polyaromatic hydrocarbons. Kadar *et al.*[116] studied the efficiences of Tenax for the preconcentration of polyaromatic hydrocarbons from standard water solutions. The method was applied to waste water samples from an aluminium plant. In this method the water samples were passed through the Tenax column at the rate of about 5 ml min^{-1}. Residual water was removed from the Tenax by passing nitrogen gas through the column. The Tenax material was then transferred to a Soxhlet apparatus. Polyaromatic hydrocarbons end other organic compounds were extracted by reflux for 4 h with 35 ml of acetone. The Tenax can then be dried and reused. Preliminary experiments indicate that the extraction time can be reduced to 10–15 min by the use of ultrasonic extraction. Residual water is removed from the acetone extract by passing the solution through a small column of anhydrous sodium sulphate and washing with 5 ml of acetone. The extract is evaporated on a water bath at 50 °C in a stream of nitrogen to a volume of approximately 0.5 ml.

The residual solution was transferred to a thin layer plate. The time required for the development is 30–35 min. A single development of the plate is sufficient for separating polyaromatic hydrocarbons from other compounds such as paraffins, naphthenes, acids, and phenols. The spots were located visually under a UV lamp and the areas containing aromatic compounds were marked. The appropriate portions of the thin layer were scraped into a glass flask. Polyaromatic hydrocarbons were extracted by vigorous shaking for 2 h with 5 ml of chloroform. Extraction of polyaromatic hydrocarbons from the thin layer material is a critical step in the procedure. Extraction into approximately 1 ml of chloroform by treatment in an ultrasonic bath for about 10 min was shown to be the best procedure. The internal standard was added to the chloroform in advance so that the samples can be analysed directly after centrifugation. After filtration or centrifugation, the solution may be used for spectrophotometric determination of total polyaromatic hydrocarbons and gas chromatographic determination of individual polyaromatic hydrocarbon components.

The conditions for the gas chromatography were as follows:

Column: glass capillary, 25 m × 0.28 mm coated with SE-30,

Detection by flame ionization detector.

Carrier gas: helium at a flow rate of 2 ml min^{-1}.

Injector temperature: 300 °C.

Detector temperature: 300 °C.

Volume injected: 1 μl.

Column temperature: initially 100 °C held for 4 min before programme starts.

Programming: 3 °C min^{-1}.

Final temperature: 260 °C.

Kadar *et al.*[116] found that the separation of polyaromatic hydrocarbons from water on the Tenax column was the most critical step in the analytical procedure (Fig. 18). Unsatisfactory results are often due to neglect of details at this step. The most important operating parameters are the height of the Tenax column and the flow rate of the water sample

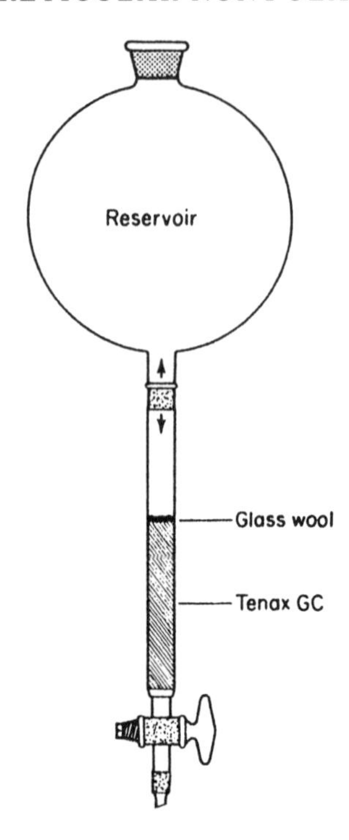

Fig.18 Sorption column with Tenax bed. From Kadar *et al.*[116]

through the column. In this work, the height of the column was set at 10 cm without any appreciable decrease in polyaromatic hydrocarbon recovery. The optimum flow rate was 8–10 ml min^{-1} for a Tenax column of 10 cm height and 13 mm diameter. The polyaromatic hydrocarbon values obtained by gas chromatography are presented in Table 60.

As indicated, excellent agreement is obtained at the 100 ng ml^{-1} level between the method and a much more time consuming liquid–liquid extraction technique. Table 60 shows the overall recovery of polyaromatic hydrocarbons from water at the 100 and the 10 ng ml^{-1} levels. At the 10 ng ml^{-1} level the recovery decreases to 70–90 per cent and the spread of the analytical results increases. At or below the 10 ng ml^{-1} level, the amount of polyaromatic hydrocarbons recovered may be increased by using a larger volume of water for instance 5 or 10 litres.

In Table 61 are shown results obtained for waste water sample from an aluminium plant. Parallel determinations were carried out on samples of 500 ml each. The standard deviation for the series is approximately 1 μg $litre^{-1}$.

Unused Tenax may contain small amounts of low molecular weight

Table 60 PAH yield from water samples.

Hydrocarbon	Amount (μg^{-1})	Liquid–liquid extraction[a] (μg^{-1})	Kadar method[116] undiluted sample		Kadar method[116] 10 times diluted sample	
			μg^{-1}	±s.d.[b]	μg^{-1}	±s.d.
Anthracene	11.0	10.3	10.2	0.5	0.8	0.1
Pyrene	32.0	30.8	30.2	0.7	2.6	0.1
Chrysene	43.0	41.3	40.3	0.6	3.2	0.3
3-Methylcholanthrene	12.0	10.9	11.0	0.3	0.9	0.1
Total	98.0	93.3	91.7		7.5	
Total yield		95	94		77	

[a] Glass capillary column g.c. method.
[b] All standard deviations (μg litre^{-1}) are based on 4 parallel results.
From Kadar *et al.*[116] with permission.

Table 61 Typical results for waste water from an aluminium plant.

Substance	Sample 1		Sample 2	
	Run 1 (μg litre^{-1})	Run 2 (μg litre^{-1})	Run 1 (μg litre^{-1})	Run 2 (μg litre^{-1})
Biphenyl	3.2	3.6	<1	<1
Fluorene/fluorenone	6.3	8.4	2.9	3.2
Phenanthrene	16.9	23.1	14.2	14.0
Anthracene	2.8	2.8	1.1	1.2
Fluoranthene	20.8	18.9	10.8	12.4
Pyrene	15.3	12.7	5.6	6.0
Benzo(a)fluorene	3.2	3.4	1.6	1.5
Benzo(b)fluorene	2.8	3.0	1.3	1.3
Benzdiphenylensulphide	3.4	3.5	2.0	1.7
Benzo(a)anthracene	2.5	2.8	5.6	5.5
Chrysene/triphenylene	5.8	6.0	15.6	16.0
Benzo(b,k)fluoranthene	6.8	6.8	38.1	38.0
Benzo(e)pyrene	2.6	2.7	16.2	16.4
Benzo(a)pyrene	1.3	1.5	7.0	7.4
Dibenzoanthracene	3.4	4.3	8.2	8.0
Anthanthrene	>1	>1	3.2	3.2
Coronene	>1	>1	1.9	2.0
Dibenzopyrene	>1	>1	4.0	4.3

From Kadar *et al.*[116] with permission.

components which can interfere with polyaromatic hydrocarbon components of the group chrysene, benzofluoranthene, benzo(*e*)pyrene on a thin layer chromatographic plate. It is therefore recommended that unused Tenax be extracted with acetone overnight. The acetone extract of unused Tenax gives rise to fluorescent spots on the thin layer plate with a retention factor of about 0.4. Since several polyaromatic hydrocarbons components have about this retention factor, these artefacts will interfere in the thin layer procedure. Judging from their intensity, this interference will be small at the 100 ng ml^{-1} level but may become dominant at lower concentrations. These artefacts will also interfere in the gas chromatographic analysis.

When analysing industrial waste water, it is advisable to protect the Tenax column against irreversible contamination. This can be achieved by using a short Celite 545 prefilter. The thin layer plate must be preconditioned in chloroform before use. This preconditioning is particularly important when the polyaromatic hydrocarbon level is very low, 10 ng ml^{-1} or less. One dimensional thin layer chromatography will give a good separation of semi-polar polyaromatic hydrocarbon components from non-polar paraffins and naphthenes which will follow the solvent front, and from the more polar acids and phenols which will remain at the bottom of the plate. Figure 19 shows a thin layer chromatogram of four polyaromatic hydrocarbon components. Under ultraviolet light the chromatographic spots may serve as a visual indication of the amount of polyaromatic hydrocarbon. In Fig. 19 it is noticeable that the total area of the chromatographic spots on the left hand side

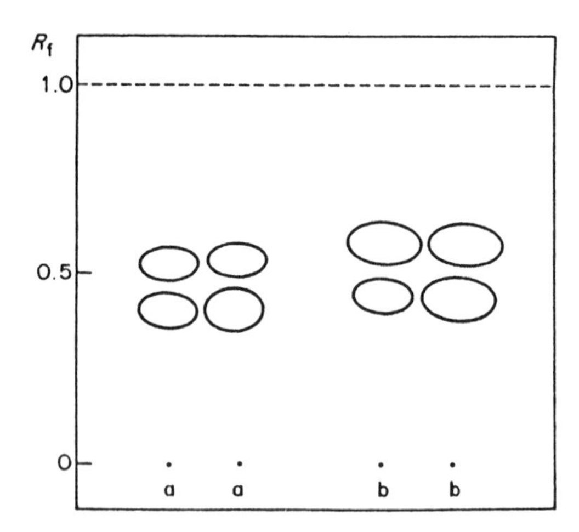

Fig. 19 TLC of PAH on Kiesegel 60 F254 (Merck); (a) sorbed on Tenax and extracted into acetone (initial concentration corresponds to 25 μg of PAH per spot); (b) initial acetone solution 25 μg of PAH per spot. From Kadar *et al.*[116] with permission.

(a) of the plate is less than the area to the right (b). Both sets of chromatographic spots correspond to the same initial amount of polyaromatic hydrocarbons (50 μg). The spots to the left, however, correspond to polyaromatic hydrocarbon components which have passed through the analytical procedure and thus have suffered a loss of about 10 per cent. This loss is visible on the plate.

Polychlorinated biphenyls and organochlorine and organophosphorus insecticides. Leoni *et al.*[117] observed that in the extractive preconcentration of organochlorine insecticides and PCBs from surface and coastal waters in the presence of other pollutants such as oil, surface active substance etc., the results obtained with an absorption column of Tenax Celite are equivalent to those obtained with the continuous liquid–liquid extraction technique. For natural waters that contain solids in suspension that absorb pesticides, it may be necessary to filter the water before extraction with Tenax and then to extract the suspended solids separately. Analyses of river and estuarine sea waters, filtered before extraction, showed the effectiveness of Tenax and the extracts obtained for the pesticide analysis prove to be much less contaminated by interfering substances than the corresponding extracts obtained by the liquid–liquid technique. Leoni *et al.*[117] showed that for the extraction of organic micropollutants such as pesticides and aromatic polycyclic hydrocarbons from waters, the recoveries of these substances from-unpolluted waters (mineral and potable waters) when added at the level of 1 μg litre^{-1} averaged 90 per cent.

Water samples were passed through the peristaltic pump into the absorption column at a flow rate of about 3 litre h^{-1}. When the absorption was completed the pesticides were eluted with three 10 ml volumes of diethyl ether, in such a way that the solvent also passed through the section of hose through which the water reached the column. Finally, the diethyl ether was dried over anhydrous sodium sulphate. The water container was washed with light petroleum to remove pesticide adsorbed on the glass and this solution, after concentration, was added to the column eluate. For the analysis of naturally polluted water the mixed diethyl ether and light petroleum extract was evaporated, the residue dissolved in light petroleum and the solution purified by partitioning with acetonitrile saturated with light petroleum.[118,119] The resulting solution was evaporated just to dryness, the residue dissolved in 1 ml of *n*-hexane and insecticides and polychlorobiphenyls were separated into four fractions by deactivated silica gel microcolumn chromatography (silica gel type Grace 950, 60–200 mesh[120]). The various eluates from the silica gel were then analysed by gas chromatography.[121] In order to evaluate the effectiveness of extraction from natural waters with the Tenax Celite column, the samples were also extracted simultaneously by the liquid–liquid technique.

In the adsorption with Tenax alone satisfactory results were obtained, while in the presence of mineral oil a considerable proportion of the organophosphorus pesticides (particularly malathion and parathion-methyl) was not adsorbed and was recovered in the filtered water. This drawback can be overcome by adding a layer of Celite 545 which, in order to prevent blocking of the column, is mixed with silanized glass wool plugs. A number of analyses of surface and estuarine sea waters were carried out by this modified Tenax column and simultaneously by the liquid–liquid extraction technique. To some of the samples taken, standard mixtures of pesticides were also added, each at the level of 1 μg litre^{-1} (i.e. in concentration from 13 to 500 times higher than that usually found in the waters analysed). One recovery trial also specifically concerned polychlorobiphenyls. The results obtained in these tests show that the two extraction methods, when applied to surface waters that were not filtered before extraction, yielded very similar results for many insecticides, with the exception of compounds of the DDT series, for which discordant results were frequently obtained.

Leoni *et al.*[117,122] conclude that the extraction of insecticide from waters by adsorption on Tenax, yields results equivalent to those by the liquid–liquid procedure when applied to mineral, potable, and surface waters that completely or almost completely lack solid matter in suspension. For waters that contain suspended solids that can adsorb some insecticides in considerable amounts, the results of the two methods are equivalent only if the water has previously been filtered. In these instances, therefore, the analysis will involve filtered water as well as the residue of filtration.

Miscellaneous organic compounds. Other applications of Tenax include the preconcentration of chlorinated and organophosphorus insecticides,[123,127] *p*-dichlorobenzene, hexachloro-1,3-butadiene, 2-chloronaphthalene,[124] organohalides,[125] trace organic compounds in ground water,[126] gasoline,[128] phenols,[129] alkyl phthalates,[129] nitrobenzene,[130] 2,4-dinitrophenol,[130] and traces of organic substances in rain-water.[131]

3.2 Metal cations

3.2.1 XAD-2 and XAD-4 non-polar resins

Non-saline waters Mackey[132] studied the adsorption of copper, zinc, iron, and magnesium ions on both these resins. Resolution was enhanced and time saved by using a multi-channel non-dispersive purpose-built atomic fluorescence detector in batch experiments. Both resins can adsorb appreciable amounts of iron, copper, and zinc ions and XAD-2 adsorbs large amounts of copper and zinc relative to its surface area. Adsorption of copper and zinc is independent of pH and, on XAD-1, independent of the amount of the other ion present. EDTA complexes are not adsorbed and a number of electrically charged metal–organic species can rapidly and completely remove iron, copper, and zinc from the resins.

Sakai and Mori[133] preconcentrated cobalt with *N*-(dithiocarboxy)sarcosine and Amberlite XAD-4 resin. Cobalt reacted with *N*-(dithiocarboxy)sarcosine to form a 1:3 cobalt *N*-(dithiocarboxy) sarcosine complex which was stable in 4 M hydrochloric acid. The complex so formed was adsorbed on a column of Amberlite XAD-4 copolymer from acid solution and eluted with 10 ml of a 1:1:3 v/v, mixture of 1.0 M ammonia solution (pH 9), 0.1 M EDTA, and methanol. The absorbance of the eluted chelate was determined at 320 nm. The extinction coefficient was 21 500 litres $mol^{-1} cm^{-1}$. Interferences were eliminated by the addition of EDTA after chelation of cobalt. The copper complex with *N*-(dithiocarboxy)sarcosine was partly adsorbed because of the slow rate of decomposition by EDTA. Most chelated copper could be eluted with hydrochloric acid and any co-eluted with the cobalt chelate could be decomposed by heating. Cobalt enrichment factors of at least 100 were obtained and the method could be applied to determination of cobalt at the ng ml^{-1} level.

Metal ions have been preconcentrated by complexation with sodium bis (2-hydroxyethyl)dithiocarbamate and sorption on XAD-4 resin.[134] Hydroxy groups present in sodium bis(2-hydroxyethyl)dithiocarbamate caused its metal complexes to be soluble in water at low concentrations. A procedure is described for the concentration of metal ions from solution on to XAD-4 resin, using sodium bis(2-hydroxylethyl)dithiocarbamate. Recovery of metals from an artificial sea-water spiked with metal salts is reported for solution pH varying from 3.0 to 10-0. Recovery rates approximated 100 per cent with lower recovery rates at extreme pH values. Selectivity could be achieved by pH adjustment and the use of masking agents such as cyanide or ethylenediaminetetraacetic acid. Advantages of the method included high concentration factors, high sample throughput, formation of soluble metal complexes, and easy elution of adsorbed metals.

Brajter *et al.*[135] modified XAD-2 resin by adsorbing Pyrocatcchol Violet on to its hydrophobic surface. They demonstrated that of iron(III), cobalt(II), nickel(II), copper(II), indium(III), lead(II), and bismith(III), only indium(III) and lead(II) were firmly retained at the appropriate pH and they used this fact to preconcentrate these two elements from potable water prior to their determination by atomic absorption spectrometry. Hiraide[136] demonstrated that indium treated XAD-2 resin retained heavy metals complexed with humic acid but not inorganic cations and anions, EDTA complexes or colloidal, hydrated ferric oxides. The sorbed heavy metal–humic acid complexes could be desorbed from the indium treated resin with aqueous nitric acid and they utilized this observation as the basis of a method for the preconcentration and determination of the humic acid complexed cations and anions.

Sea-water Fujita and Iwashima[137] preconcentrated mercury compounds in sea-water by first forming the diethyldithiocarbamate and then concentrating this on XAD-2 resin. The resin was eluted with methanol/3 M hydrochloric acid; the organic mercury was extracted with benzene and then back-extracted with cysteine solutions. The organic mercury in the cysteine solution and the total mercury adsorbed on the resin were determined by flameless atomic absorption spectrometry. The method was applied to determinations of mercury levels in sea-water in and around the Japanese archipelago. The lower limit of detection in sea-water is 0.1 ng $litre^{-1}$ for organic mercury, using 801 samples.

Hirose and Sugimura[138] investigated the speciation of plutonium in sea-water using adsorption of plutonium(IV)–xylenol orange and plutonium–arsenazo(III) complexes on the macroreticular synthetic resin XAD-2. Xylenol orange was selective for plutonium(IV) and arsenazo(III) for total plutonium. Plutonium levels were determined by α-ray spectrometry.

Isshiki and Nakayama[146] preconcentrated cobalt in sea-water by complexing it with 4-(2-thiazolylazo)resorcinol to form a water soluble inert complex and passing this solution down a column of XAD-4 resin. The complex was then eluted with methanol/chloroform, the eluate digested with nitric and perchloric acids, and cobalt determined by graphite furnace atomic absorption spectrometry.

3.3 Organometallic compounds

3.3.1 Texax GC

Sea-water – tin Brinckmann and co-workers[139] used a gas chromographic method with or without hydride derivatization for determining volatile organotin compounds (e.g. tetramethyltin), in sea-water. For non-volatile organotin compounds a direct liquid chromatographic method was used. This system employs a 'Tenax-GC' polymeric sorbent in an automatic purge and trap (P/T) sampler coupled to a conventional glass column gas chromatograph equipped with a flame photometric detector. Figure 20 is a schematic of the P/T–GC–FPD assembly with typical operation conditions. Flame conditions in the FPD were tuned

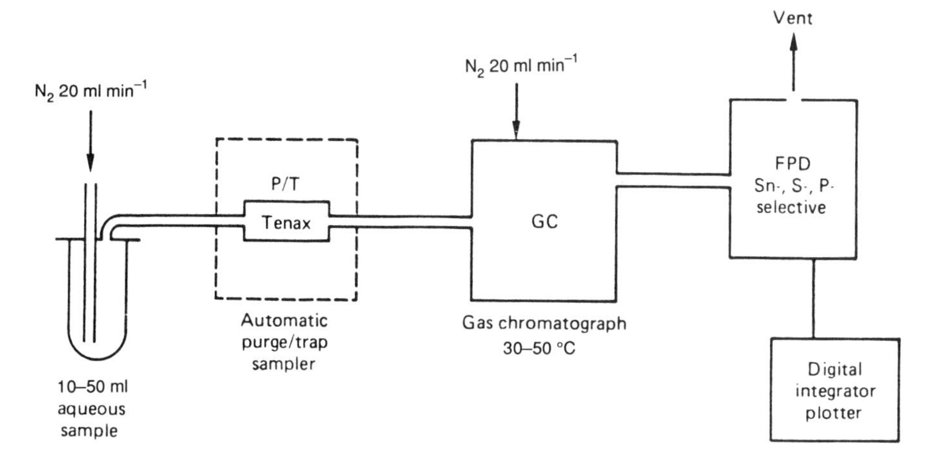

Fig. 20 The purge/trap GC–FPD system and operating conditions. From Brinckman.[139]

to permit maximum response to SnH emission in a H-rich plasma, as detected through narrow band-pass interference filters (610 ± 5 nm).[140] Two modes of analysis were used:

(1) volatile stannanes were trapped directly from sparged 10–50 ml water samples with no pretreatment; and

(2) volatilized tin species were trapped from the same or replicate water samples following rapid injection of aqueous excess sodium borohydride solution directly into the P/T sparging vessel immediately prior to beginning the P/T cycle.[141]

Jackson *et al.*[142] devised trace speciation methods capable of ensuring detection of tin species along with appropriate preconcentration and derivatization without loss, decomposition, or alteration of their basic molecular features. They describe the development of a system empolying a Tenax GC filled purge and trap sampler, which collects and concentrates volatile organotins from water samples (and species volatilized by hydrodization with sodium borohydride), coupled automatically to a gas chromatograph equipped with a commercial flame photometric detector modified for tin-specific detection.[143-145]

References

1. Daignault, S. A., Noot, K. K., Williams, D. T., and Huck, P. M. (1988). *Water Research*, *22*, 803.
2. Wigilius, B., Boren, H., Carlberg, G. E., Grimvall, A., Lundgren, B. V., and Savenhead, R. (1987), *Journal of Chromatography*, *391*, 169.

3. Dietrich, A. D., Millington, D. S., and Seo, Y. H. (1988). *Journal of Chromatography*, *436*, 229.
4. Stephan, S. F. and Smith, J. F. (1977). *Water Research*, *11*, 339.
5. Chang, R. C. and Fritz, J. S. (1978). *Talanta*, *25*, 659.
6. Ryan, J. P. and Fritz, J. S. (1978). *Journal of Chromatographic Science*, *16*, 448.
7. Tateda, A. and Fritz, J. S. (1978). *Journal of Chromatography*, *152*, 329.
8. Junk, G. A., Richard, J. J., Grieser, D., Witiak, J. L., Arguello, M. D., Viel, R., Svec, J. S., Fritz, J. S., and Calder, G. V. (1974). *Journal of Chromatography*, *99*, 745.
9. Fielding, M., Gibson, T. M., James, H. A., McLoughlin, L., and Steel, C. P. (1981). In *Water Research Centre Technical Report TR 159. Organic micropollutants in drinking water.*) February.
10. Stephen, S. F., Smith, J. F., Flego, U., and Renkens, J. (1978) *Water Research*, *12*, 447.
11. Benoit, F. M., Label, G. L., and Williams, D. T. (1979). *International Journal of Analytical Chemistry*, *6*, 277.
12. McNeil, E. E., Olson, R., Miles, W. F., and Rajabalee, R. J. M. (1977). *Journal of Chromatography*, *132*, 277.
13. Nickless, G. and Jones. P. (1978). *Journal of Chromatography*, *156*, 87.
14. Musty, P. R. and Nickless, G. (1974). *Journal of Chromatography*, *89*, 185.
15. Musty, P. R. and Nickless, G. (1976). *Journal of Chromatography*, *120*, 369.
16. Heatley, N. G. and Page, E. J. (1952). *Water Sanitation*, *3*, 46.
17. Rosen, M. J. and Goldsmith, H. A. (1972). In *Systematic Analysis of Surface Active Agents*, 2nd edn, Wiley, New York.
18. Patterson, S. J., Hunt, E. C., and Tucker, K. B. E. (1966). *Journal of Proceedings of Institute of Sewage Purification*, *190*.
19. Jones, P. and Nickless, G. (1978). *Journal of Chromatography*, *156*, 99.
20. Nadeau, H. G. and Waszeciak, P. H. (1967). In *Non-ionic Surfactants*, Marcel Dekker, New York, p. 906.
21. Junk, G. A. and Richard, J. J. (1988). *Analytical Chemistry*, *60*, 451.
22. Paschal, D. C., Bicknell, R., and Dresbach, D. (1977). *Analytical Chemistry*, *49*, 1551.
23. Junk, G. A. (1974). *Journal of Chromatography*, *99*, 745.
24. Coburn, J. A., Valdamanis, J. A., and Chau, A. S. Y. (1977). *Journal of Association of Official Analytical Chemists*, *60*, 224.
25. Berkane, K., Caissie, G. E., and Mallet, V. N. (1977). In *Proceedings of Symposium on Fenitrothion*, NRC (Canada) Assoc. Comm. Sci. Crit. Environ. Quality, Report 16073 NRCC, Ottawa, 95.
26. Daughton, C. G., Crosby, D. G., Garnos, R. L., and Ssieh, J. (1976). *Journal of Agricultural and Food Chemistry*, *24*, 236.
27. Mallet, V. N., Brun, G. L., MacDonald, R. N., and Berkane, K. (1978). *Journal of Chromatography*, *160*, 81.
28. Berkane, K., Caissie, G. E., and Mallet, V. N. (1977). *Journal of Chromatography*, *139*, 386.

29. Zakrevsky, J. G. and Mallet, V. N. (1977). *Journal of Chromatography*, *132*, 315.
30. Volpe, G. G. and Mallet, V. N. (1980). *International Journal of Environmental Analytical Chemistry*, *8*, 291.
31. Puijker, L. M., Veenendaal, G., Jaonssen, H. M. J., and Griepin, B. (1981). Fresenius Zeitschrift fur Analytische Chemie, *306*, 1.
32. Frobe, Z., Drevenkar, V., Stengl, B., and Stefanac, Z. (1988). *Analytica Chimica Acta*, *206*, 299.
33. Akhing, B. and Jensen, S. (1970). *Analytical Chemistry*, *42*, 1483.
34. Richards, J. F. and Fritz, J. S. (1974). *Talanta*, *21*, 91.
35. Levesque, D. and Mallet, V. N. (1983). *International Journal of Environmental Analysis*, *16*, 139.
36. Narange, A. S. and Eadon, G. (1982). *International Journal of Environmental Analytical Chemistry*, *11*, 167.
37. Harris, R. L., Huggett, R. J., and Stone, H. D. (1980). *Analytical Chemistry*, *52*, 779.
38. Sundaram, K. M. S., Szeto, S. Y., and Hindle, R. (1979). *Journal of Chromatography*, *177*, 20.
39. Ishibashi, M. and Suzuki, M. (1988). *Journal of Chromatography*, *456*, 382.
40. Rees, G. A. V. and Au, L. (1979). *Bulletin of Environmental Toxicology*, *22*, 561.
41. Norsdsij, A., Van Beveren, J., and Brandt, A. (1984). H_2O, *17*, 242.
42. Coburn, J. A., Valdeamis, I. A., and Chau, A. S. Y. (1977). *Journal of the Association of Official Analytical Chemists*, *60*, 224.
43. Elder, D. (1976). *Marine Pollution Bulletin*, *7*, 63.
44. Harvey, G. R. (1973). In *Report US Environment Protection Agency*, EPA-R2-73-177, 32 pp.
45. Niederschulte, U. and Ballschmiter, K. (1974). *Zeit Analytical Chemistry*, *269*, 360.
46. Richard, J. J. and Fritz, J. S. (1974). *Talanta*, *21*, 91.
47. Junk, G. A., Richard, J.J., Griser, M. D., Witiak, D., Witiak, J. L., Arguello, M. D., Vick, R., Svec, H. J., Fritz, J. S., and Calder, G. V. (1974). *Journal of Chromatography*, *99*, 745.
48. Chriswell, C. D., Ericson, R. L., Junk, G. A., Lee, K. W., Fritz, J. S., and Svec, H. J., (1977). *Journal of the American Waterworks Association*, *69*, 669.
49. Musty, P. R. and Nickless, G. (1976). *Journal of Chromatography*, *120*, 369.
50. Lawrence, J. and Tosine, H. M. (1976). *Environmental Science and Technology*, *10*, 381.
51. Le'Bel, K. and Williams, D. T. (1980). *Bulletin of Environmental Contamination and Toxicology*, *24*, 397.
52. Sjostrom, L., Radestrom, R., Carlberg, G. E., and Kringstad, A. (1985). *Chemosphere*, *14*, 1107.
53. Renberg, L. (1978). *Analytical Chemistry*, *50*, 1836.
54. Mantoura, R. F. C. and Riley, J. P. (1975). *Analytica Chimica Acta*, *76*, 97.

55. Anon, (1976). *Chemical Engineering News*, *54*, 35.
56. Kissinger, L. D. (1979). In *Report No IS-T-845*, US National Technical Information Service, Springfield, Virginia, Va., 161 pp.
57. Kissinger, L.D. and Fritz, L. S. (1976). *Journal of the American Water Works Association*, *68*, 435.
58. Renberg, L. (1978). *Analytical Chemistry*, *50*, 1836.
59. Daughton, C. G., Crosby, D. G., Garmos, R. L., and Hseih, P. P. H. (1976). *Journal of Agricultural and Food Chemistry*, *24*, 236.
60. Prater, W. A., Simmons, H. C., and Mancy, K. M. (1980). *Analytical Letters*, *13*, 205.
61. Junk, G. A., Richard, J. J., Grieser, M. D., Willink, D., Witlak, M.D., Argnello, M. D., Vick, R., Svec, H. J., Fritz, J. S., and Calder, G. V. T. (1974). *Journal of Chromatography*, *99*, 745.
62. Richard, J. J. and Fritz, J. S., (1980). *Journal of Chromatographic Science*, *18*, 35.
63. Noll, K. E. and Gounaris, A. (1988). *Water Research*, *22*, 815.
64. Chau, Y. K. and Fox, M. E. (1971). *Journal of Chromatographic Science*, *9*, 271.
65. Oliver, B. G. and Bothen, K. D. (1980). *Analytical Chemistry*, *52*, 2066.
66. Cabrident, R. and Solika, A. (1975). *TSM L'Eau*, *285*, July.
67. Burnham, A. K., Calder, G. V., Fritz, J. S., Junk, G. A., Svec, H. J., and Williams, R. (1972). *Analytical Chemistry*, *44*, 139.
68. Junk, G. A., Richard, J. J., Grieser, M. D., Witiak, D., Witiak, J. L., Arguello, M. D., Vick, R., Svec, H. J., Fritz, J. S., and Calder, G. V. (1974). *Journal of Chromatography*, *99*, 745.
69. Fritz, J. S., (1975). *Industrial Engineering Produce Development Research*, *14*, 95.
70. Leenheer, J. A. and Huffman, E. W. D. (1976). *Journal of Research, US Geological Survey*, *4*, 737.
71. Tateda, A. and Fritz, J. S. (1978). *Journal of Chromatography*, *152*, 382.
72. Stephen, S. F., Smith, J. F., Flegoil, and Rember, J. (1978). *Water Research*, *12*, 447.
73. Chang, R. C. and Fritz, J. S. (1978). *Talanta*, *25*, 659.
74. Ryan, J. P. and Fritz, J. S. (1978). *Journal Chromatographic Science*, *16*, 488.
75. Stepan, S. I. and Smith, J. F. (1977). *Water Research*, *11*, 339.
76. Ishangir, L. M. and Samuelson, O. (1978). *Analytica Chimica Acta*, *100*, 53.
77. More, R. A. and Karasek, F. W. (1984). *International Journal of Environmental Analytical Chemistry*, *17*, 187.
78. James, H. A., Steel, C. P., and Wilson, I. (1981). *Journal of Chromatography*, *208*, 89.
79. Webb, R. G. (1975). National Technical Information Service, Springfield, Virginia, PB 245674, 22 pp.
80. Chudaba, J., Chlebkova, E., and Tucek, F. (1977). *Vodni Hospodarstvi*, *27B*, 236.
81. Chang, R. C. and Fritz, J. S. (1978). *Talanta*, *25*, 659.
82. Fox, M. E. (1977). *Journal of the Fisheries Research Board of Canada*, *34*, 798.

83. Ishiwatari, R., Hamara, H., and Machihara, T. (1980). *Water Research*, *14*, 1257.
84. De Groat, R. (1979). *H_2O*, *12*, 333.
85. Burnham, A. K., Calder, G. V., Fritz, J. S., Junk, G. A., Svec, H. J., and Willis, R. (1972). *Analytical Chemistry*, *44*, 139.
86. De Groat, R. (1979). *H_2O*, *12*, 333.
87. Noordsij, A. (1979). *H_2O*, *12*, 167.
88. Riley, J. P. and Taylor, D., (1969). *Analytica Chimica Acta*, *46*, 307.
89. Burnham, A. K., Calder, G. V., Fritz, J. S., Junk, G. A., Svec, H. J., and Willis, R. (1972). *Analytical Chemistry*, *44*, 139.
90. Leenheer, J. A. and Huffman, E. W. D. (1976). *Journal of Research of the Geological Survey*, *4*, 737.
91. Gustafson, R. L. and Paleos, J. (1971). Interactions responsible for the selective adsorption of organics on organic surfaces in *Organic Compounds in Aquatic Environments*, eds S. J. Faust and J. V. Hunter, Marcel Dekker, New York, pp. 213–37.
92. Riley, J. P. and Taylor, D. (1969). *Analytica Chimica Acts*, *46*, 307.
93. Osterroht, C. (1972). *Kiel, Meeresforsch*, *28*, 48.
94. Osterroht, C. (1974). *Journal of Chromatography*, *101*, 289.
95. Musty, P. R. and Nickless, G. (1974). *Journal of Chromatography*, *89*, 185.
96. Elder, D. (1976). *Marine Pollution Bulletin*, *7*, 63.
97. Harvey, G. R. (1973). *Report of the US Environment Protection Agency* EDA-R2-73-177.
98. Picer, N. and Picer, M. (1980). *Journal of Chromatography*, *193*, 357.
99. Holden, A. V. and Marsden, K. (1969). *Journal of Chromatography*, *44*, 481.
100. Snyder, D. E. and Reinert, R. E. (1971). *Bulletin of Environmental Contamination and Toxicology*, *6*, 385.
101. Picer, M. and Abel, M. (1978). *Journal of Chromatography*, *150*, 1191.
102. Burnham, A. K. (1972). *Analytical Chemistry*, *44*, 139.
103. Ahling, B. and Jenson, S. (1970). *Analytical Chemistry*, *42*, 1483.
104. Harvey, G. R. (1972). In *Absorption of chlorinated hydrocarbons from seawater by a crosslinked polymer*. Woods Hole Oceanographic Institute, Woods Hole, Massachusetts, Published manuscript.
105. Gomez-Belinchon, J. I., Grimalt, J. O., and Albaiges, J. (1988). *Environmental Science and Technology*, *22*, 677.
106. Daughton, C. G., Crosby, D. C., Garnas, R. L., and Hseih. D. O. H. (1976). *Journal of Agriculture and Food Chemistry*, *24*, 236.
107. Dawson, R., Riley. J. P., and Tennant, R. H. (1976). *Marine Chemistry*, *4*, 83.
108. Shinohara, R., Kido, A., Okomoto, Y., and Takeshita, R. (1983). *Journal of Chromatography*, *256*, 81.
109. Junk, G. A., Richard, J. J., Greeser, M. D. (1974). *Journal of Chromatography*, *99*, 745.
110. Stuermer, D. H. and Harvey, G. R. (1974). *Nature (London)*, *250*, 480.
111. Stuermer, D. H. and Harvey, G. R. (1977). *Deep Sea Research*, *24*, 303.
112. Mantoura, R. F. C. and Riley, J. P. (1975). *Analytical Chimica Acta*, *76*, 97.

113. Wun, C. K., Walker, R. W., and Litsky, W. (1976), *Water Research*, *10*, 955.
114. Wun, C. K., Walker, R. W., and Litsky, W. (1978). *Health Laboratory Science*, *15*, 67.
115. Przyazny, A. (1988). *Journal of Chromatography*, *346*, 61.
116. Kadar, R., Nagy, K., and Fremtad, D. (1980). *Talanta*, *27*, 227.
117. Leoni, V., Pucetti, G., and Grella, A. J. (1975). *Journal of Chromatography*, *106*, 119.
118. Leoni, V. (1971). *Journal of Chromatography*, *62*, 63.
119. Johnston, L. V. (1965). *Journal of Association of Official Analytical Chemists*, *48*, 668.
120. Claeys, R. R. and Inman, R. D. (1974). *Journal of Association of Official Analytical Chemists*, *57*, 399.
121. Leoni, V. and Pucetti, G. (1969). *Journal of Chromatography*, *43*, 388.
122. Leoni, V., Pucetti, G., Columbo, R. J., and Oviddo, O. (1976). *Journal of Chromatography*, *125*, 399.
123. Agostiano, A., Caselli, M., and Provenzano, M. A. (1983). *Water Air and Soil Pollution*, *19*, 309.
124. Parkow, J. F. and Isabelle, L. M. (1982). *Journal of Chromatography*, *237*, 25.
125. Sekerka, L. and Lechner, J. F. (1982). *International Journal of Environmental Analytical Chemistry*, *11*, 43.
126. Jankow, J. F., Isabelle, L. M., Hewetsen, J. P., and Cherry, J. A. (1985). *Ground water*, *23*, 775.
127. Tator, V. and Popl, M. (1985). *Fresenius Zeitschrift Fur Analytische Chemie*, *322*, 419.
128. Belkin, F. and Hable, F. A. (1988). *Bulletin of Environmental Contamination and Toxicology*, *40*, 244.
129. Pankow, J. F., Ligocki, M. P., Rosen, M. E., Isabelle, L. M., and Hart, K. M. (1988). *Analytical Chemistry*, *60*, 40.
130. Patil, S. C. (1988). *Analytical Letters*, *21*, 1397.
131. Pankow, J. F., Isabelle, L. M., and Asher, W. E. (1984). *Environmental Science and Technology*, *18*, 310.
132. Mackey, D. J. (1982). *Journal of Chromatography*, *236*, 81.
133. Sakai, Y. and Mori, N. (1986). *Talanta*, *33*, 161.
134. King, J. N. and Fritz, J. S. (1985). *Analytical Chemistry*, *57*, 1016.
135. Brajter, K., Olbrych-Sleszynska, E., and Stastiewicz, M. (1988). *Talanta*, *35*, 65.
136. Hiraide, M., Arima, Y., and Mizurka, A. (1987). *Analytica Chimica Acta*, *200*, 171.
137. Fujita, M. and Iwashima, K. (1981). *Environmental Science and Technology*, *15*, 929.
138. Hirose, K. and Sigimura, Y. J. (1985). *Radioanalytical and Nuclear Chemistry Articles*, *92*, 363.
139. Brinckmann, F. E. (1981). In *Trace Metals in Seawater, Proceedings of a Nato Advanced Research Institute on Trace Metals in Seawater*, 30/3–3/4/81, Sicily, Italy, eds C. S. Wong *et al.*, Plenum Press, New York.

140. Aue, W. A. and Flinn, C. S. (1977). *Journal of Chromatography*, *142*, 145.
141. Jackson, J. A., Blair, W. R., Brinckmann, F. E., and Iverson, W. P. (1982). *Environmental Science and Technology*, *16*, 110.
142. Jackson, J. A., Blair, W. R., Brinckmann, F. E., and Iveson, W. P. (1982). *Environmental Science and Technology*, *16*, 111.
143. Aue, W. A. and Flinn, G. C. (1977). *Journal of Chromatography*, *142*, 145.
144. Huey, C., Brinckmann, F. E., Grim, S., and Iveson, W. P., (1974) In *Proceedings of the International Conference on the Transport of Persistent Chemicals in Equatic Ecosystems*, eds A. S. W. Freilas, D. J. Kushner, and D. S. U. Quadri, National Research Council of Canada, Ottawa, Canada, pp 11–73–11–78.
145. Nelson, J. D., Blair, W., and Brinckmann, F. E. (1973). *Applied Microbiology*, *26*, 321.
146. Isshiki, K. and Nakayama, E. (1978). *Analytical Chemistry*, *59*, 291.
147. Musty, P.R. and Nickless, G. (1974). *Journal of Chromatography*, *89*, 185.

4

CATION EXCHANGE RESINS

In addition to the non-polar macroreticular Rohm and Hass and Amberlite XAD resins (XAD-2 and XAD-4) discussed in Chapter 3 a wide range of polar resins exist which are very useful for the preconcentration of anionic and cationic species and have some applications in the preconcentration of ionic organic substances (see Table 62). Intermediate and highly polar types of resins are commonly referred to as ion exchange resins. These may be subdivided into catonic types (cation exchange resins) which are discussed in this chapter and anionic types (anion exchange resins) which are discussed in Chapter 5. Cationic exchange resins carry a negative charge and this reacts with positively charged metallic ions (cations) or cationic organic species. Anionic ion exchange resins carry a positive charge and this reacts with negatively charged anions or anionic organic species.

Strong acid cation exchange resins manufactured by the sulphonation of polystyrene or polydivinyl benzenes undergo the following reaction with cations:

$$\text{Resin } SO_3^- H^+ + M^+X^- \rightarrow \text{Resin } SO_3^- M^+ + H^+ + X^-$$

Weak acid cation exchange resins manufactured, e.g. by the polymerization of methacrylic acid undergo the following reaction with cations:

$$\text{Resin } COO^- H^+ + M^+ + X^1 \rightarrow \text{Resin } COO^- M^+ + H^+ + X^-$$

Some basic properties of the various types of cation exchange resins available and their suppliers are tabulated in Table 62.

Preconcentration is achieved by passing a large volume of water sample, suitably adjusted in pH and reagent composition down a small column of the resin. The adsorbed ions are then desorbed with a small volume of a suitable reagent in which the metals or metal complexes or anionic species dissolve. This preconcentrated extract can then be analysed by any suitable means.

4.1 Metal cations

4.1.1 Conventional cation exchange resins—non-saline waters

Atomic absorption spectrometry Kempster and Van Vliet[1] have described a semi-automated resin concentration method for the preconcentration of trace metals (chromium, manganese, iron, cobalt, nickel, copper,

Table 62 Properties of cation exchange resins.

Resin type	Functional group	Water content (approx)[a] ($g\,g^{-1}$ dry resin)	Exchange capacity (approx)[a] (mol equiv g^{-1} dry resin)	Packing density (approx)[a] ($g\,ml^{-1}$)	Regeneration	Washing of salt forms	Trade volumes of some commercial examples
Strong acid types	$-SO_3H$	0.7	4 at all pH values	0.8	Excess strong acid	Stable	(c) Dowex 50 Dowex 50W-X8 Dowex 50W-X4 Dowex A1 (b) Amberlite 1R-120 Amberlite GC-120 Amberlite XAD-12 (a) Zeocarb 225 Cationite KB-4P-2
Weak acid types	$-COOH$	1	9–10 at high pH	0.7	Readily regenerated	Cation slowly hydrolyses off	(b) Amberlite 1RC-50 (a) Zeocarb 226 (c) Dowex XAD-7 Dowex XAD-8

[a] Depends on grade and does not necessarily include recently developed resins available from (a) Permutit Co., London W4; (b) Rohm and Haas Co., Philadelphia, USA; (c) Dow Chemical Co., Midland, Michigan, USA.

zinc, cadmium, and lead) in potable water, prior to atomic absorption analysis. A peristaltic pump was used to control the flow of water samples through columns of a cation exchange resin (Amberlite IR-120/H form), the samples being stabilized with ascorbic acid (0.5 g $litre^{-1}$) at a pH of 2.5 during the sorption stage. A 26 channel peristaltic pump was used to pump the contents of 25 samples and one blank simultaneously through the resin columns, using 1.6 mm i.d. polyethylene for transmission lines and 0.86 mm i.d. tygon tubing in the peristaltic pump, which gave a constant flow rate of 0.42 ml min^{-1} and a flow-through time in the resin column of just under 5 min. Before use the resin columns were freed, if necessary, of any entrapped air by detaching the columns from the flow lines and aspirating deionized water rapidly through each column with a syringe. The resin columns were then cleaned by pumping 12 ml 5 M acid (redistilled AR) followed by 12 ml deionized water through the columns.

The water samples analysed were treated as follows: the samples, collected in precleaned polyethylene bottles and preserved with concentrated nitric acid (5 ml $litre^{-1}$), were filtered through a 0.45 μm pore size membrane filter to remove particulate matter. Ascorbic acid (0.5 g $litre^{-1}$) was then added, and the pH adjusted to between 2.0 and 2.5 with concentrated ammonia solution using pH indicator paper. The samples were then pumped through the cleaned resin columns (at 0.42 ml min^{-1}) to sorb the metals on to the resin. The time necessary to pump 500 ml sample volumes through the resin columns was just under 20 h. With subsequent elution of the sorbed metals to a volume of 50 ml, a concentration factor of 10× was obtained. The volume of each sample passed through the resin columns was measured to determine the exact concentration ratio.

To elute the metals from the resin columns, 25 ml 5 M hydrochloric acid followed by 12 ml of deionized water was pumped through each column, the eluate being collected in 50 ml volumetric flasks. Concentrated ammonia solution (3.5 ml) was then added to the elute in each 50 ml volumetric flask to reduce the excessive acidity, which was found to produce noisy signals in the subsequent atomic absorption analysis. The pH of the eluate after the addition of ammonia solution was found to be less than 2. The eluate was made up to 50 ml with deionized water and transferred to 50 ml polyethylene bottles until analysis by flame atomic absorption spectroscopy. The resin columns were then regenerated for the next batch of samples with 25 ml 3 M hydrochloric acid (redistilled AR).

Of the nine elements mentioned above, eight gave a recovery through the whole procedure of between 88 and 99 per cent whilst iron had a recovery of 75 per cent. Detection limits (μg $litre^{-1}$) were as follows: Cr, 3; Mn, 0.5; Co, 1; Cu, 0.5; Zn, 2; Cd, 0.1; Pb, 6. This method is useful for the preconcentration of a large number of samples, does away with the tedium characteristic of manual enrichment techniques, and gives good recovery for the nine metals tested. Difficulty was experienced in obtaining consistent results for iron but it was found that with the addition of ascorbic acid to the samples, prior to sorption on to the resin, more consistent results were obtained. Ascorbic acid serves as a complexing agent to counteract hydrolysis of iron.

Treit *et al.*[2] also linked a cation exchange resin preconcentration column (Dowex 50W-X8) directly to the nebulizer tube of an atomic absorption spectrometer in their method for the determination of free metal ions including copper. They used a miniaturized ion exchange column. The metal ion is eluted from the resin as a narrow peak, the area of which is proportional to the free metal ion concentration in the initial sample solution. The spectrophotometer is thus used as an ion selective probe. Precision and accuracy were better than 1 per cent. The method allows free metal determination in the sample and both sample volumes and measurement times are reduced.

Reagents:

Stock solution 0.100 M prepared by dissolution of pure copper wire in nitric acid.

Sodium nitrate (British Drug House).

Sodium salt of ethylenediaminetetra-acetic acid, EDTA (Fisher), analytical grade

Distilled water, passed through a mixed bed ion-exchange column and filtered through 10–20 μm sintered glass to remove particulates.

Resin column:

Slurry analytical grade 200–400 mesh Dowex 50W-X8 cation exchange resin (Dow Chemical Co.) with an exchange capacity of 5.1 mequiv g^{-1} of dry resin in water and remove the fines by decanting. Wash the resin with hydrochloric acid, sodium hydroxide, ethanol, and water and dry as described by Cantwell *et al.*[3]

Pack approximately 2 mm long resin bed into a 50 mm length of 1.5 mm i.d. Teflon tubing flared at both ends (Fig. 21). This length of tubing was required to accommodate the Cheminert and fittings (ODC). Slurry pack the resin into the tube as follows: prepare approximately 2 mm long glass frits by drilling cores from a sintered glass plate (nominal pore size 50 μm). Press one frit about 6 mm into the 15 mm i.d. Teflon tube and butt up against a 0.063 in od. Teflon retaining tube. With this inner frit in place, inject a slurry of about 0.1 ml of resin in several millilitres of water with a 10 ml glass syringe. Attach the syringe to the column tube via a Luer adaptor (LDC) and a short

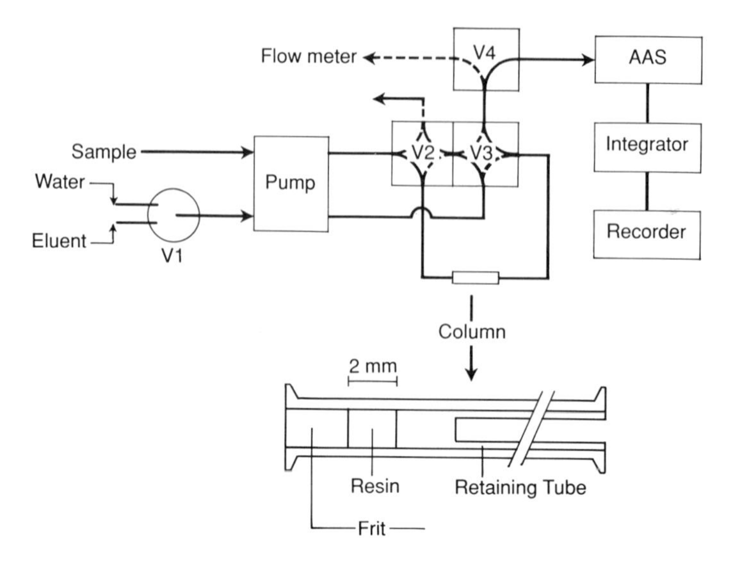

Fig. 21 Diagram of instrument system and ion exchange column; see text for details. From Treit *et al.*[2] with permission.

length of 1.5 mm i.d. Teflon tubing, using Cheminert bushings and end fittings. When the column was filled gently remove the outer 2 mm of resin and press a second 2 mm long glass frit into place. Take care not to excessively compact the resin to avoid possible flow inhibition.

Apparatus:

The instrument is shown in Fig. 21. The variable speed peristaltic pump (Minipula 2, Gilson, Villiers-le-Bel, France) was fitted with Clear Standard (PVC) tubes (Technicon Corp.). One tube pumped sample (or standard) copper containing solution and the other pumped either water or solvent, depending on the position of the Teflon rotary value V_1 (Model R6031 V6, Laboratory Data Control LDC, Riviera Beach, FL). Pump rates were measured with a burette and stopwatch by diverting the flow through the three port Teflon slider valve V_4 (CAV3031 LDC). By means of the two coupled four port Teflon slider valves V_2 and V_3 (CAV4031 LDC) the flow from either pump tube could be directed through the resin column while the other was diverted to waste. The resin column was immersed in a 30 ± 1 °C water bath.

Monitor copper concentration by flame AA spectrometry with a Model 290B spectrophotometer (Perkin Elmer Corp) or equivalent. Connect a piece of 0.3 mm i.d. Teflon tubing from valve V_4 directly to the nebulizer tubing. Use a lean acetylene–air flame along with the following operating conditions: wavelength, 324.7 nm; slit width, 0.7 nm; and meter damping, 3. The AA signal was recorded on a

Model 7101BM strip chart recorder (Hewlett Packard Corp.) and eluted copper peak areas were measured with a Model 23000–011 Autolab Minigrator (Spectra Physics Corp).

Determination of free copper ion:

Prepare all copper containing samples and standards in 0.100 M sodium nitrate. Use a flow rate of 5.5 ml min^{-1}. Pump sample solution through the 2 mm resin column via V_2 and V_3 (solid lines in Fig. 21). Conduct this resin equilibration step for 7.5 min. Divert the effluent load solution to the flow meter by valve V_4 during this step to prevent fouling of the nebulizer and burner by the high salt concentration.

After the effluent is loaded, switch V_2, V_3, and V_4 (dashed lines in Fig. 21) and backwash the resin with water for 2.5 min to remove interstitial solution and to establish a base line AA signal. Next elute copper on the resin by switching V_1 to 0.02 M EDTA, pH 3.5 eluent for 1.5 min. Detect the eluted copper as an AA peak and measure the peak area. After elution, again wash the column with water for 2.5 min to reestablish a base line. Then return valves V_2, V_3, and V_4 to their initial positions for the next sample.

A significant advantage of this technique is that the ion exchange resin is brought to equilibrium with the sample in such a way as to allow free metal determination in the unperturbed sample. Furthermore, the use of very small columns permits faster measurements with smaller sample volumes. Still smaller sample volumes may be realized by reducing resin particle size, lowering flow rates during the loading step, and decreasing the amount of resin. Use of a graphite furnace in place of a flame would also permit reduced sample volumes but would require indirect coupling to the column. The semi-automated system is especially useful for measuring free ion concentrations and complexation capacities of biological fluids, natural waters, waste water, and other aquatic samples.

The use of this technique requires 'trace conditions' of ion exchange with a relatively large and virtually identical concentration of the same major cation (e.g. sodium) in both sample and standard. For most samples this is readily achieved by adding the same concentration of 'swamping electrolyte' (e.g. sodium nitrate) to all. Other workers have devised equipment based on ion exchange enrichment of metals followed by direct desorption of the concentrated metal into an atomic absorption spectrometer.[4,5] Fang *et al.*[4] used a flow injection system with ion exchange preconcentration for the determination of trace amounts of heavy metals in natural waters. A multifunctional rotary valve, incorporating two parallel miniature ion exchange columns packed with a chelating resin, containing salicylic acid functional groups, was used for sequential sampling, injection, ion exchange, and elution. The sensitivity was increased 20 to 28-fold compared to direct aspiration of samples, for nickel, copper, lead, and cadmium at a sampling rate of 40 per hour.

Fang *et al.*[5] found that the sensitivity of the method was comparable to that of the graphite furnace technique and was cheaper and simpler. Exact timing was used for sample metering and injection, in combination with constant pumping rates. The detection limits ranged from 0.05 to 0.5 μg litre^{-1} with relative standard deviations 0.2–3.2 per cent. Samara and Kouimtzia[6] preconcentrated cobalt, chromium, copper, iron, nickel, and zinc from reactor cooling water with an Acropane resin prior to desorption with acid and determination by atomic absorption spectrometry.

Pilipenko *et al.*[7] determined the adsorption properties of five cation exchangers for copper(II), zinc(II), lead(II), manganese(II), cobalt(II), nickel(II), and cadmium(II) in the same solution at different pH values. After sorption of the cations, the exchangers were transferred to a column and eluted with hydrochloric acid for subsequent analysis by atomic absorption spectrometry. Type KU-2 and KU-23 cation exchangers were the most convenient exchangers to use, as type KU-1 and KB-4 required a long contact time and type KB-4 was also strongly affected by the pH. The effects of other cations and anions present in natural waters was negligible.

Werner[8] used Dowex 50WX-4 resin in the calcium form to adsorb cadmium and zinc from a nitriloacetic acid buffer and samples rich in humic acid prior to desorption and analysis by atomic absorption spectrometry. pH had a profound effect on the binding capacity of the resin for cadmium in humic acid rich samples. Between 1 and 3 μg litre^{-1} cadmium and zinc could be determined by this method.

Sweilek *et al.*[9] used an ion exchange column to preconcentrate copper(II) species from solutions containing citrate, glycinate, phthalate, salicylate, chloride, and fulvate. The preconcentrated metal was desorbed and determined by atomic absorption spectrometry. The method was subject to interference from cationic and neutral copper complexes as well as from filterable colloid copper–hydroxo species at higher pH values. The method would be particularly useful for determination of divalent copper in natural waters where the copper was often present in low concentrations, where cationic and neutral copper complexes were likely to be absent, and where humates and fulvates were the principal complexing agents.

Neutron activation analysis Preconcentration on a cation exchange membrane followed by neutron activation analysis has been used to determine extremely low levels of cadmium in natural waters.[10] Linstedt and Kruger[11] preconcentrated vanadium(V) from natural waters on to

Dowex 50-X8 cation exchange resin. The vanadium was then desorbed with nitric acid prior to determination in amounts down to 0.1 μg $litre^{-1}$ in the sample by neutron activation analysis. Duffy *et al.*[12] preconcentrated aluminium, calcium, potassium, magnesium, manganese, and sodium from soil pore water on to a cation exchange column, having first removed humic materials on an anion exchange column. Cations were then desorbed and determined by neutron activation analysis.

X-ray fluorescence spectrometry Levesque and Mallet[13] preconcentrated zinc, cadmium, nickel, lead, cobalt, manganese, iron, and chromium on a column of Amberlite IR-120 resin prior to desorption with sodium chloride solution and determination by X-ray fluorescence spectrometry. Ho and Lin[14] preconcentrated calcium, iron, cobalt, nickel, copper, zinc, lead, mercury, chromium, and selenium by passing the potable water sample repeatedly through a cation exchange resin. The resin was subsequently analysed by energy dispersive X-ray fluorescence spectrometry. These workers used Amberlite IR-120 paper for calcium, iron, cobalt, nickel, copper, lead, and zinc, and Amberlite LRS-400 paper for mercury, selenium, and chromium. Detection limits ranged from 0.5 μg $litre^{-1}$ (iron) to 9 μg $litre^{-1}$ (selenium).

Other analytical finishes spectrophotometric, spectrographic, ion-selective electrode, and radiochemical methods have also been employed as analytical finishes in cation preconcentration techniques, see Table 63.

4.1.2 Cation exchange resins—sea-water

Atomic absorption spectrometry Wan *et al.*[28] determined conditions for the direct preconcentration of cadmium, manganese, chromium, copper, nickel, iron, cobalt, and lead from sea-water samples using a two column Amberlite XAD-7 resin system. Low breakthrough volumes in the presence of humic materials necessitated their prior removal at a pH of 1–2 prior to preconcentration of the trace metals on a second column of XAD-7 at pH 8. Metals were subsequently desorbed from the second column with 1 per cent nitric acid by means of a precolumn of XAD-7. The final effluent for measurement by graphite furnace atomic absorption spectrometry is readily matrix matched and permits use of the standard calibration curve procedure. Preconcentration factors of 40 were obtained by this procedure permitting the analysis of coastal sea-waters for the eight elements mentioned earlier.

Dehairs *et al.*[29] give details of a procedure for the determination of barium in sea-water, involving separation of barium from manganese by collection on a cation exchange resin in the ammonium form and extraction of the barium from the resin into nitric acid for determination by graphite furnace atomic absorption spectrometry.

Table 63 Preconcentration on cation exchange resins.

Metals	Type of water	Resin	Medium	Eluting agent	Detection limit	Analytical finish	Ref.
Mn	River	Amberlite CG-120	Acetic acid	M-ammonium chloride	3 μg	Spectrophotometric	15
				M-ammonium thiocyanate			
Zn	Natural	Amberlite CG-120	Acetic acid	Ammonium Chloride	2 μg $litre^{-1}$	Spectrophotometric	16
Cr^{III}	Natural sea-water	Cationite KB-4P-2	Sodium chloride	–	–	Spectrophotometric	17
Cu^{II}	Natural	Cationic resin	–	–	–	Spectrophotometric	20
Co, Ni, Cu, Zn, Pb, Sn	Natural	KU-2	None	Hydrochloric acid	–	Spectrographic	18

Cu	Natural	–	–	–	–	Ion-selective electrode atomic absorption spectrometry	19
^{22}Na	Natural	Cationic resin (H form)	–	Hydrochloric acid	–	β–γ counting	21,22
^{137}Cs	Natural	Amberlite IR 120 CP (H form)	–	–	–	Radiochemical	23,25,26
^{90}Sr	Natural	KP-P4-2	–	–	–	Radiochemical	24–26
^{137}Cs	Natural	KU-2	–	–	–	Radiochemical	24–26
^{89}Sr,	Natural	Cationic resin	–	–	5 femto Ci litre^{-1}		25,26
^{226}Ra-226	Natural	Cationic	–	–	0.03 Ci litre^{-1}	Liquid scintillation counting	27

4.1.3 Other cation exchange materials

Ammonium molybdophosphate

Non-saline waters. Caesium-137 has been preconcentrated from 10–100 litre samples of river water, by passage through a bed of ammonium molybdophosphate which is subsequently examined by γ-ray spectrometry[30] or by measurement of its β activity.[31-37] Down to 0.3 pCi litre^{-1} caesium-137 activity can be measured by these procedures.

Sea-water. Dutton[38] has described a procedure for the determination of caesium-137 in water. This procedure comprises a simple one-step separation of the radio-caesium from the sample using ammonium dodecamolybdophosphate or potassium cobaltihexacyanoferrate and ^{137}Cs and ^{134}Cs are measured by γ-ray counting of the dried adsorbent with a NaI (TI) crystal coupled to a γ-ray spectrometer. Level of ^{137}Cs activity down to about 1 pCi litre^{-1} can be determined in sea-water and lake, rain, and river waters without sophisticated chemical processing.

Ammonium hexacyanocobalt ferrate

Sea-water. Atomic absorption spectrometry has been used to determine caesium in sea-water.[39,40] The method uses preliminary chromatographic separation on a strong cation exchange resin, ammonium hexacyanocobalt ferrate, followed by electrothermal atomic absorption spectrometry. The procedure is convenient, versatile, and reliable, although decomposition products from the exchanger, namely iron and cobalt can cause interference. Caesium is fully retained by a chromatographic column of ammonium hexacyanocobalt ferrate and can then be recovered by dissolution of the ammonium hexacyanocobalt ferrate in hot 12 M sulphuric acid. The resin is stable in strong acid solutions.

A further method for the determination of caesium isotopes in saline waters[41] is based on the high selectivity of ammonium hexacyanocobalt ferrate for caesium. The sample (100–500 ml) is made 1 M in hydrochloric acid and 0.5 M in hydrofluoric acid, then stirred for 5–10 min with 100 mg of the ferrocyanide. When the material has settled, it is collected on a filter (pore size 0.45 μm), washed with water, drained, dried under an infra-red lamp, covered with plastic film and β-counted for ^{137}Cs. If ^{131}Cs is also present, the γ-spectrometric method of Yamamoto[42] must be used. Caesium can be determined at levels down to 10 pCi litre^{-1}.

Zinc hexacyanoferrate–agar-agar

Non-saline waters. Caletka *et al.*[43] preconcentrated ^{137}Cs from natural water samples on zinc hexacyanoferrate bonded on agar-agar gel. The method achieved a yield of at least 95 per cent and a preconcentration factor of up to 5000.

4.1.4 Liquid chelating exchangers

Non-saline waters Bhattacharyya and Das[44] used a beta-diketo derivative of Versatic-10, to extract mercury selectively from industrial effluents. Mercury(II) in the separated organic phase was reduced to the elemental state with stannous chloride prior to its determination by cold vapour atomic absorption spectrometry.

4.2 Organics

4.2.1 Cation exchange resins

Non-saline waters In determinations of organic impurities in waste water, e.g. in industrial effluents, treatment with a cation exchange resin in free acid form is often used to remove metal cations and cationic organic solutes such as amino acids. Possible losses of compound such as fatty acids and aromatic compounds have been discussed.[44–49]

Jahangir and Samuelson[50,51] have discussed the sorption of cyclohexane derivatives with a hydroxyl or carbonyl group separated by one or two methylene groups from aqueous media by these resins. They found that these compounds adsorbed more strongly than the corresponding aromatic compounds both on sulphonated styrene–devinylbenzene cation exchange resins and on non-ionic styrene–divinylbenzene resins. These observations and the lower temperature dependence observed for the cyclohexane derivatives indicate that hydrophobic interactions have a marked influence on the adsorption.

Kaczvinsky *et al.*[52] give details of the use of a cation exchange resin (polystyrene–divinylbenzene type) for preconcentration of trace organic compounds from water. Neutral and acid compounds can be eluted from the resin by washing with methanol and ethyl ether; basic compounds can be eluted with the same solvents after treatment of the resin with ammonia gas. Over 50 organic bases were recovered from water at 1 mg litre^{-1} to 50 μg litre^{-1} levels at rates of over 85 per cent for most compounds studied. Nielen *et al.*[53] used a strongly acidic cation exchange resin for on-line preconcentration of polar anilines in water. The method could be automated and a detection limit for the nine anilines examined corresponding to 0.02–0.5 μg litre^{-1} was obtained in river water samples.

Capacity factors have been evaluated[54] for the adsorption of organic solutes on Amberlite XAD-8 resin. Some applications of cation exchange resins to the preconcentration of organics from water samples are listed in Table 64.

Sea-water Although free amino acids are present only at very low concentrations in oceanic waters, their importance in most biological systems has led to an inordinate amount of effort toward their determination in sea-water. A sensitive, simple, and easily automated method of analysis, the colorimetric ninhydrin reaction, has been known in biochemical research for many years. In order for the method to be useful in sea-water, the amino acids had to be concentrated. This concentration was usually achieved by some form of ion-exchange.[66]

Ligand exchange was used as a concentrating mechanism by Clark *et al.*[67] followed by TLC for the final separation. The formation of 2,4-dinitro-1-fluorobenzene derivatives, followed by solvent extraction of these derivatives and circular TLC was suggested by Palmork.[68] Ligand exchange has been a favoured method for the concentration of amino acids from solution because of its selectivity.[69] Gardner,[70] isolated free amino acids at the 20 nmol litre^{-1} level in from as little as 5 ml of sample, by cation exchange, and measured concentrations on a sensitive amino acid analyser equipped with a fluorimetric detector.

The classical work of Dawson and Pritchard[71] on the determination of α-amino acids in sea-water uses a standard amino acid analyser modified to incorporate a fluorimetric detection system. In this method the sea-water samples are desalinated on cation exchange resins and concentrated prior to analysis. The output of the fluorimeter is fed through a potential divider and low-pass filter to a compensation recorder. Dawson and Pritchard[71] point out that all procedures used for concentrating organic components from sea-waters, however mild and uncontaminating, are open to criticism, simply because of the ignorance as to the nature of these components in sea-water. It is, for instance, feasible that during the process of desalting on ion exchange resins under weakly acidic conditions metal chelates dissociate and thereby larger quantities of 'free' components are released and analysed.

4.3 Organometallic compounds

4.3.1 Cation exchange resins

Non-saline waters—tin Neubert and Andreas[72] preconcentrated tri- and di-butyltin species from water on a cation exchange column, then desorbed them with a small volume of diethyl ether–hydrogen chloride prior to their conversion to methyltin species and gas chromatographic analysis.

Table 64 Applications of intermediate polarity (XAD-7, XAD-8) and high polarity (XAD-12) cation exchange resins to preconcentrations of organics from waters.

Substance	Resin	Method of desorption	Analytical finish	Detection limit for organic matter	Type of water sample	Ref.
Mischellaneous	XAD-7	–	–	mg litre^{-1}	Well water	55
Miscellaneous	XAD-8	CS_2 extraction	GLC–MS	0.1 μg litre^{-1}	Surface waters	56
Miscellaneous	XAD-7	Solvent extraction	GLC	–	Natural	57
Miscellaneous	XAD-7	Diethyl ether/ methanol extraction	GLC	–	Natural	58
Fenitrothion	XAD-7	Solvent extraction	GLC	0.5 ng	Natural	59
Aminocarb	XAD-7	Solvent extraction	GLC with N–P detector	–	Natural	60
Urea (as indophenol)	XAD-7	Aqueous extraction	–	–	Natural	61
Alkylethoxylated sulphates	XAD-7	Methanolic hydrochloric acid	Derivativization to alkylbromides–GLC	–	Waste and surface waters	62
Phenols	Ambersorb XE-340	Diethyl ether or dichloromethane	GLC	0.5 μg litre^{-1}	Natural	63
Nitrosamines	Ambersorb XE-340	Solvent extraction	GLC	–	Potable	64
Alkylbenzene sulphonates	XAD-8	Methanol	^{13}C NMR	–	Sewage effluents	65

Sea-water-arsenic Persson and Irgum[73] determined sub-ppm levels of dimethyl arsinate by preconcentrating the organoarsenic compound on a strong cation exchange resin (Dowex AG 50 W-WB). By optimizing the elution parameters, dimethyl arsinate can be separated from other arsenicals and sample components, such as Group I and II metals, which can interfere in the final determination. Graphite furnace atomic absorption spectrometry was used as a sensitive and specific detector for arsenic. The described technique allows dimethyl arsinate to be determined in a sample (20 ml) containing a 10^5-fold excess of inorganic arsenic with a detection limit of 0.02 μg $litre^{-1}$ arsenic. Good recoveries were obtained from the artificial sea-waters, even at the 0.05 μg $litre^{-1}$ level, but for natural sea-water samples the recoveries were lower (74–85 per cent). This effect could be attributed to organic sample components that eluted from the column together with dimethyl arsinate.

References

1. Kempster, B. L. and Van Vliet, H. R. (1978). *Water South Africa*, *4*, 125.
2. Treit, J., Nielson, J. S., Kratochvil B., and Cantwell, F. F. (1983). *Analytical Chemistry*, *55*, 1650.
3. Cantwell, F. F., Nielsen, J. S., and Hrudy, S. E. (1982). *Analytical Chemistry*, *54*, 1498.
4. Fang, Z., Xu, S., and Zhang, S. (1984). *Analytica Chemica Acta*, *164*, 41.
5. Fang, Z., Ruzicka, J., and Hansen, E. H. (1984). *Analytica Chimica Acta*, *164*, 23.
6. Samara, C. and Kouimtzia, T. A. (1987). *Chemosphere*, *16*, 405.
7. Pilipenko, A. T., Safronova, V. G., and Zakrevskaya, L. V. (1987). *Soviet Journal of Water Chemistry and Technology*, *9*, 74.
8. Werner, J. (1987). *Science of the Total Environment*, *62*, 281.
9. Sweilek, J. A., Lucyk, D., Kratochvil, B., and Cantwell, F. F. (1987). *Analytical Chemistry*, *59*, 586.
10. Mark, H. B., Eisner, U., and Rothschafler, J. M. (1969). *Environmental Science and Technology*, *3*, 165.
11. Linstedt, K. D. and Kruger, D. (1970). *Analytical Chemistry*, *42*, 113.
12. Duffy, S. J., Hay, G. W., Micklethwaite, R. K., and Van Loon, G. W. (1988). *Science of the Total Environment*, *76*, 203.
13. Levesque, D. and Mallet, V. N. (1983). *International Journal of Environmental Analyses*, *16*, 139.
14. Ho, J. S. Y. and Lin, P. C. L. (1982). *International Laboratory*, *12*, 44.
15. Matsui, H. (1974). *Analytica Chimica Acta*, *62*, 216.
16. Matsui, H. (1973). *Analytica Chimica Acta*, *66*, 143.
17. Suranova, Z. P., Oleinck, G. M., and Morozov, A. A. (1969). *Īzvestia Vyssh. ucheb Zavob. Khim. Tekhnology*, *12*, 149. Ref: *Zhur Khim*. 19GD (14) Abstract No. 14G82 (1969).
18. Marshall, M. A. and Mottola, A. (1985). *Analytical Chemistry*, *57*, 729.

19. Sweilch, J. A., Lucyk, D., Kratochvil, B., and Cantwell, F. F. (1987). *Analytical Chemistry*, *59*, 586.
20. Yoshimura, K., Nigo, S., and Tarutani, T., (1982). *Talanta*, *29*, 173.
21. Yasulenis, R. Y., Luyanas, V. Y., and Kekite, V. P. (1972). *Soviet Radiochemistry*, *14*, 673.
22. Burden, B. A. (1968). *Analyst* (*London*), *93*, 715.
23. Stewart, M. L., Pendleton, R. E., and Lords, J. L. (1972). *International Journal of Applied Radiation and Isotopes*, *23*, 345.
24. Kapustin, V. K., Egorov, A. I., and Leonov, V. V. (1981). *Soviet Journal of Water Chemistry and Technology*, *3*, 119.
25. Sevagacnik, M. and Paljk, S. (1969). *Zeitung Analytical Chemistry*, *244*, 306.
26. Sevagacnik, M. and Paljk, S. (1969). *Zeitung Analytical Chemistry*, *244*, 375.
27. Higachi, H., Uesugi, M., Satoh, K., Ohashie, N., and Norguchi, N. (1984). *Analytical Chemistry*, *56*, 761.
28. Wan, C. C., Chaing, S., and Corsini, A. (1985). *Analytical Chemistry*, *57*, 719.
29. Dehairs, F., De Bandt, M., Baeyens, W., Van den Winkel, F., and Hoenig, M. (1987). *Analytica Chimica Acta*, *196*, 33.
30. Mundschenk, H., (1974) *Deutsche Gewasserkundliche Mittalunger*, *18*, 72.
31. Rehak, W. and Ubl, G. (1970). In *Report Staatliche. Zeitrate. Suer Stranlenschutz*, S25-1/70.
32. Stewart, M. L., Pendleton, R. C., and Lords, J. (1972). *Journal of Applied Radiation and Isotopes*, *23*, 345.
33. Senegacnik, M. and Paljk, S. (1969). *Zeitung Analytical Chemistry*, *244*, 306.
34. Senegacnik, M. and Paljk, S. (1969). *Zeitung Analytical Chemistry*, *244*, 375.
35. Kleberer, K. and Stuerzer, U. (1970). *Gas-u-Wasserfach*, *111*, 29.
36. Tereda, K., Hayakawa, H., Savada, K., and Kibu, T. (1970). *Talanta*, *17*, 955.
37. Kapustin, V. K., Egorov, A. I., and Leonov, V. L. (1981). *Soviet Journal of Water Chemistry and Technology*, *3*, 119.
38. Dutton, J. W. R. (1970). In *Report of the Fisheries and Radiobiological Laboratory*, FRL 6, Ministry of Agriculture, Fisheries and Food, UK.
39. Frigieri, P., Trucco, R., Ciaccolini, I., and Pampurini, G. (1980). *Analyst* (*London*), *105*, 651.
40. Frigieri, P., Trucco. R., Ciaccolini, I., and Pampurini, G. (1980). *Analyst* (*London*), *105*, 651.
41. Janzer, V. J. (1973). *Journal of Research of the US Geological Survey*, *1*, 113.
42. Yamamoto, O. (1967). *Analytical Abstracts*, *14*, 6669.
43. Caletka, R., Munster, H., and Krivan, O. (1987). *Fresenius Zeitchrift Fur Analytische Chemie*, *327*, 19.
44. Bhattacharyya, S. S. and Das, A. K. (1988). *Atomic Spectroscopy*, *9*, 68.
45. Kingston, H. and Pella, P. A. (1981). *Analytical Chemistry*, *53*, 223.
46. Samuelson, S. V. (1942). *Kem*, *Tidskr*, *54*, 170.

47. Vaisman, A. and Yampolskaya, M. M. (1950). *Zavod. Lab.*, *156*, 621.
48. Seki, T. (1954). *Journal of Chemical Society of Japan Pure Chemistry Section*, *75*, 1297.
49. Nomura, N., Hirski, S., Yamada, M., and Shiho, D. (1971). *Journal of Chromatography*, *59*, 373.
50. Jahangir, L. M. and Samuelson, O. (1976). *Analytica Chimica Acta*, *85*, 103.
51. Jahangir, L. M. and Samuelson, O. (1978). *Analytica Chimica Acta*, *100*, 53.
52. Kaczvinsky, J. R. Saitoh, K., and Fritz, J. S. (1983). *Analytical Chemistry*, *55*, 1210.
53. Nielen, M. W. F., Frei, R. W., and Brinkman, U. A. T. (1984). *Journal of Chromatography*, *317*, 557.
54. Thurman, E. M., Malcolm, R. L., and Aiken, G. R. (1978). *Analytical Chemistry*, *50*, 775.
55. Burnham, A. K., Calder, G. V., Fritz, J. S., Junk, G. A., Svec, H. J., and Willis, R. (1972). *Analytical Chemistry*, *44*, 139.
56. De Groot, R. (1979). *H_2O*, *12*, 333.
57. Stepan, S. F. and Smith, J. F. (1977). *Water Research*, *11*, 339.
58. Stephan, S. F., Smith, J. F., Flego, Ul., and Renkers, J. (1978). *Water Research*, *12*, 447.
59. Volpe, G. G. and Mallet, N. N. (1980). *International Journal of Environmental Analytical Chemistry*, *8*, 291.
60. Levesque, D. and Mallet, V. N. (1983). *International Journal of Environmental Analysis*, *16*, 139.
61. Moreno, P., Sanchez, E., Pons, A., and Palon, A. (1986). *Analytical Chemistry*, *58*, 585.
62. Neubecker, T. A. (1985). *Environmental Science and Technology*, *19*, 1232.
63. Cavelier, C., Gilber, M., Vivien, L, and Lamblin, P. (1984). *Revue Francais des Sciences de l'eau*, *3*, 19.
64. Kimoto, W. I., Dooley, C. J., Carre, J., and Fiddler, W. (1981). *Water Research*, *15*, 1099.
65. Thurman, E. M., Willoughby, T., Barber, L. B., and Thorn, K. A. (1987). *Analytical Chemistry*, *59*, 1798.
66. Semenov, A. D., Ivleva, I. N., and Datsko, V. G. (1964). *Trudy Komiss, Anal. Khim. Akad, Nauk. SSSR*, *13*, 162.
67. Clark, M. E., Jackson, G. A., and North, W. J. (1972). *Limnology and Oceanography*, *17*, 749.
68. Palmork, K. H. (1963). *Acta Chimica Scandanavia*, *17*, 1456.
69. Siegal, A. and Degens, E. T. (1966). *Science*, *151*, 1098.
70. Gardner, W. S. (1975). *Marine Chemistry*, *6*, 15.
71. Dawson, R. and Pritchard, R. G. (1978). *Marine Chemistry*, *6*, 27.
72. Neubert, G. and Andreas, H. (1976). *Zeitung Analytical Chemistry*, *280*, 31.
73. Perssen, J. and Irgum, K. (1982). *Analytica Chimica Acta*, *138*, 111.

5
ANION EXCHANGE RESINS

Strong base anion exchange resins are manufactured by chloromethylation of sulphonated polystyrene followed by reaction with a tertiary amine:

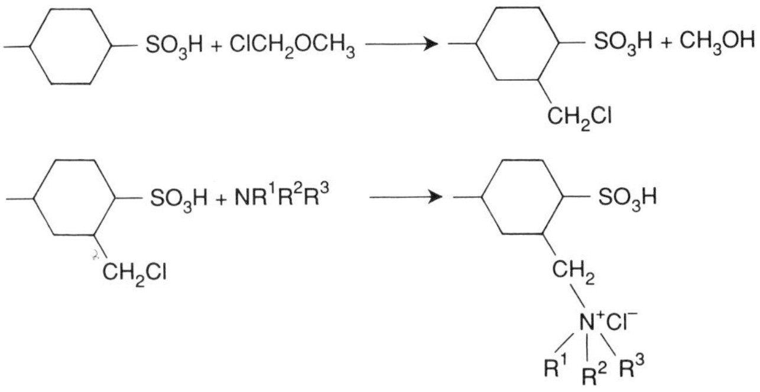

$$-C_6H_4-SO_3H + ClCH_2OCH_3 \longrightarrow -C_6H_3(CH_2Cl)-SO_3H + CH_3OH$$

$$-C_6H_3(CH_2Cl)-SO_3H + NR^1R^2R^3 \longrightarrow -C_6H_3(CH_2N^+R^1R^2R^3\,Cl^-)-SO_3H$$

They undergo the following reaction with anions:

$$\text{Resin}-CH_2N^+(R_1R_2R_3)Cl^- + M^+ + X^- \rightarrow \text{Resin}-CH_2-N^+(R_1R_2R_3)X^- + Cl^- + M^+$$

or

$$\text{Resin}-CH_2-N^+(R_1R_2R_3)Cl^- + MX^- \rightarrow \text{Resin}-CH_2-N^+(R_1R_2R_3)MX^- + Cl^-$$

where MX^- is a metal containing anion.

Weak base anion exchange resins are manufactured by chloromethylation of sulphonated polystyrene followed by reaction with a primary or secondary amine:

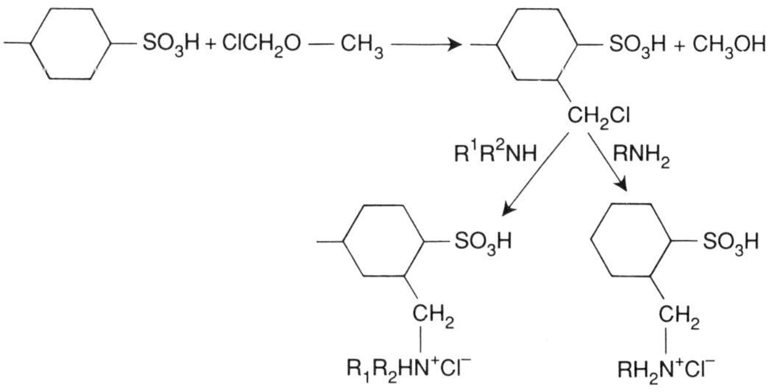

$$-C_6H_4-SO_3H + ClCH_2O-CH_3 \longrightarrow -C_6H_3(CH_2Cl)-SO_3H + CH_3OH$$

$$-C_6H_3(CH_2Cl)-SO_3H \xrightarrow{R^1R^2NH} -C_6H_3(CH_2-R_1R_2HN^+Cl^-)-SO_3H$$

$$-C_6H_3(CH_2Cl)-SO_3H \xrightarrow{RNH_2} C_6H_3(CH_2-RH_2N^+Cl^-)-SO_3H$$

They undergo the following reaction with anions:

e.g.

$$\text{Resin} - \underset{R^1 R^2 H}{\overset{}{N^+}}Cl^- + X^- + M^+ \rightarrow \text{Resin} - \underset{R^1 R^2 H}{N^+}X^- + Cl^- + M^+$$

or

$$\text{Resin} - \underset{R^1 R^2 H}{N^+}Cl^- + MX^- \rightarrow \text{Resin} - \underset{R^1 R^2 H}{N^+}MX^- + Cl^-$$

where MX^- is a metal containing anion.

Some basic properties of the various types of anion exchange resins and suppliers are tabulated in Table 65.

5.1 Simple anions—non-saline waters

5.1.1 Chromate

Various ion exchange resins have been used to preconcentrate chromate ions including Dowex AG1-X4 anion exchange resin, Amberlite La-1 liquid anion exchanger,[1,2] and Dowex 501W-X 4 cation change resin.

Parkow *et al.*[3] have described a procedure for the differential analysis of traces of chromium(VI) (chromate) and chromium(III) in natural waters. The sample is filtered, acidified, and divided into three portions, one of which is left untreated while the others are passed through a cation exchange resin and an anion exchange resin, respectively. The three aliquots are then treated with nitric acid, evaporated, and analysed by atomic absorption to give the concentrations of cationic, anionic, and non-ionic chromium. The concentration of chromium(III) is probably closely related to the sum of the cationic and non-ionic fractions and the concentration of chromium(VI) corresponds to the anion portion.

Equipment:

Analyses for chromium were performed on a Perkin Elmer 303 Atomic Absorption spectrophotometer, with a Honeywell Electronix 194 recorder.

Materials:

Potassium dichromate and all other chemicals and buffer solutions used were reagent grade.

Anion exchange resin, AG1-X4 (Cl-form, 100–200 mesh) and cation exchange resin 50W-X4 (Na form, 100–200 mesh): analytical grade, Bio-Rad Laboratories.

Deionized water, distilled from an all Pyrex still and used for dilution and for the rinsing of (1 M HNO_3 washed) glassware.

Table 65 Properties of anion exchange resins.

Resin type	Functional group	Water content (approx)[a] ($g\,g^{-1}$ dry resin)	Exchange capacity (approx)[a] (mol equiv g^{-1} dry resin)	Packing density (approx)[a] ($g\,ml^{-1}$)	Regeneration	Washing of salt forms	Trade names of same commercial examples
Strong base types	Quarternary ammonium $-CH_2NR_3 + Cl^-$	1	4 at all pH values	0.7	Excess strong base	Stable	(c) Dowex 1 Dowex 2 Dowex AG1-X2 Dowex AG1-X8 Dowex AG1-X4 (b) Amberlyst P1-27 Amberlite IRA-400 Amberlite GC-400 Amberlite IRA-410 (a) Deacidite FF Lewatite M5080 Biorad AG1 Biorad 140-AG1-X2 Biorad X8
Weak base types	Secondary or primary amine $-CH_2\text{-}N^+HR_2Cl^-$ or $-CH_2\text{-}N^+H_2R\,Cl^-$	0.3	4 at low pH values	0.7	Readily regenerated with sodium carbonate	Anion slowly hydrolyses	(b) Amberlite IR-45 (a) Deacidite G.

[a]Depends on grade and does not necessarily include recently developed resins available from (a) Permitit Co., London W4; (b) Rohm and Haas Co., Philadelphia, USA; (c) Dow Chemical Co., Midland, Michigan, USA.

Glassfibre filter papers (diameter 70 mm S&S No. 25) used to remove suspended matter from river water samples: Schleicher and Schuell Corporation.

Membrane filters: (0.45 μm pore size), Millipore Corporation.

Procedure:

Collect samples in (1 M nitric acid washed, doubly deionized water rinsed) polyethylene containers. Rinse the containers (4 litres in volume) several times with the water to be analysed. Filter a 3 litre water sample through glass fibre filters. After filtration, divide the 3-litre sample into equal portions of 1 litre each. Pass one aliquot through a 4.0 $\times$ 0.5 cm column of anion exchange resin, one through a similar column of cation exchange resin, and leave one untreated. Add 10 ml of 1.0 M nitric acid to each litre aliquot and then, slowly (over 4 h) evaporate (without boiling) to a volume of 10 ml. Analyse by atomic absorption spectrometry using the Cr line at 357.9 nm and the method of standard additions using a reducing flame (yellow) to maximize the chromium signal. The concentrations of total, anionic, cationic, and non-ionic chromium in the original 3-litre sample are obtained by difference. The precisions of the analyses for total, anionic, cationic, and non-ionic chromium were respectively $\pm$14, $\pm$ 20, $\pm$20, and $\pm$25 per cent relative.

Dankow[4] acidified water samples containing chromate to pH 5 and passed them upwards through an anion exchange resin, so that the chromate is adsorbed in a narrow zone at the lower end of the resin bed. The chromate is eluted rapidly with small volumes of an acidic reductant solution which reacts with chromate on the column to form trivalent chromium during elution, thus producing very high concentration factors.

5.1.2 Phosphate

In one procedure[5,6] phosphate is adsorbed on to anion exchange resin. Orthophosphate is quantitatively adsorbed by a AG Dowex 1-X8 anion exchange resin, then eluted and reacted with an acid molybdate reagent for estimation. Arsenic and organic phosphorus compounds did not interfere with the estimation of orthophosphate while polyphosphates did interfere if present in equal or greater amounts than orthophosphate. It is concluded that the use of the anion exchange technique results in a more valid estimate than direct reaction with the acid molybdate reagent.

5.1.3 Sulphide

Sulphide has been preconcentrated on a column of Amberlite IRA 400 anion exchange resin.[7] The sulphide is removed from the column with

4 M sodium hydroxide and determined spectrophotometrically by the *N*, *N'*-dimethyl-*p*-phenylene diamine method. Down to 0.1 μg litre^{-1} sulphide can be determined by this procedure.

5.1.4 Borate

Jun *et al.*[8] preconcentrated boron from natural waters as its complex with chromotropic acid and octyltrimethyl ammonium chloride in an anion exchange column (TSK gel, IC Anion Pw). The eluted concentrate was analysed by high performance liquid chromatography.

5.2 Anionic metal species

Simple cationic metal ions do not react with, i.e. are not retained by, anion exchange resins. However, if the metal cation M^+ is first reacted with a reagent with which it forms a negatively charged anionic complex then the resulting negatively charged metal containing ions are retained on an anionic exchange column. Thus cadmium(II) forms a soluble anionic complex upon reaction with potassium cyanide:

$$CdCl_2 + 4\,KCN \rightarrow K_2[Cd(CN)_4]^{2-}$$

This complex is retained on a column of strong base anionic exchange resin which, for convenience, is represented as follows:

$$2\,\text{Resin}\,CH_2R_3N^+Cl^- + [Cd(CN)_4]^{2-} \rightarrow \text{Resin}\,CH_2R_3N^+]_2\,[Cd(CN)4]^{2-} + 2\,Cl^-$$

The cadmium complex can then be dissolved off the column with a small volume of aqueous acetic acid to procedure the acetate form of the resin and cadmium acetate:

$$[\text{Resin} - CH_2R_3N^+]_2(Cd(CN)_4)^{2-} + 4CH_3COOH \rightarrow 2\,\text{Resin} - CH_2R_3N^+[OOCCH_3]^- + Cd(OOCCH_3)_2 + 4\,HCN$$

High preconcentration factors can be achieved by such techniques, some further examples of which are now discussed.

5.2.1 Non-saline waters

Atomic absorption spectrometry Korkische and Sario[9] have described a procedure for the determination of cadmium(II), copper(II), and lead(II) in natural non-saline waters. The method involved acidification with hydrobromic acid; filtration of the water sample; addition of ascorbic acid; ion exchange using the strongly basic anion exchange resin, Dowex 1-X8, which adsorbs cadmium, copper, and lead as anionic bromide complexes, of the type $H_2M\ Br_4$ – (where M is metal); elution with nitric acid; and determination by atomic absorption spectrometry. When the methanolic hydrobromic acid solution was used, the following sensitivities for 1 per cent absorption were obtained; Cd = 0.033 mg $litre^{-1}$; Cu = 0.04 mg $litre^{-1}$; Pb = 0.28 mg $litre^{-1}$.

In the determination of extremely small quantities of cadmium, copper, and lead, it is necessary to run a reagent blank through the whole procedure (starting with the addition of concentrated hydrobromic acid and finally to deduct its concentration of cadmium, copper, and lead from those contents measured in the water samples.

To investigate the effect of the volume of water sample on the recovery and accuracy of the determinations of cadmium, copper, and lead, Korkische and Sario[9] analysed varying volumes of potable water for these three elements. The volume of water has no effect provided that it does not exceed 500 ml. If larger volumes of the water sample are passed through the anion exchange column, the adsorption of copper and lead decreases with increasing volume. This is because of the salt effect, i.e. the displacement of the anionic bromo complexes of these metals by sulphate and bromide ion contained in the water (bromides of calcium and magnesium are formed from the carbonates on acidification of the water sample with hydrobromic acid). With respect to the accuracy of the determinations of copper and lead, the volume has practically no influence (in the range of 0.1–0.5 litre). It is impossible, however, to determine cadmium when only 100–500 ml of the sample is used, in this case, at least 1 litre of the sample has to be passed through the anion exchange column to obtain an eluate which contains sufficient cadmium to be measurable with some accuracy. Therefore, relatively large volumes of water have to be processed in order to obtain reliable cadmium values.

The main constituents of natural waters i.e. calcium, magnesium, the alkali metals, and iron, are not retained on Dowex 1 from 0.15 M hydrobromic acid. The effect of various metal ions on the anion exchange separation of cadmium, copper, and lead was studied with respect to some trace elements. It was found that 1 mg amounts of most common elements did not affect the recoveries of cadmium, copper, and lead. Among these foreign metal ions, only zinc may be present in natural waters to any larger extent but even then it would not interfere because zinc is not retained by the resin from 0.15 M hydrobromic acid.

The ligand 8-hydroxyquinoline-5-sulphonic acid forms anionic complexes with cobalt(II), zinc(II), cadmium(II), and lead(II). A technique has been described[10] for the separation and preconcentration of these metals prior to measurement by graphite furnace atomic absorption spectrometry. At optimum conditions, all four metals are quantitatively retained as their negatively charged 8-hydroxyquinoline-5-sulphonic acid complexes by the column. Zinc, cadmium, and lead(II) ions are completely eluted with 11 ml or less of 2 M nitric acid; cobalt(II) is totally removed by 12 M hydrochloric acid. All four anionic complexes can be left on the column for 7 days and still be quantitatively (99 per cent) recovered.

Reagents:

8-Hydroxyquinoline-5-sulphonic acid (Aldrich Chemical Company), recrystallized from water, 200×10^{-3} M stock solution is prepared using a minimum amount of sodium hydroxide to solubilize the compound.

pH 8 buffer, prepare by mixing 50 ml of 0.1 M Tris with 20 ml of 0.1 M hydrochloric acid and diluting the mixture to 100 ml.

Anion exchange resin (Bio-Rad 140–1242 AG I-X2, 100–200 mesh) used: in the OH^- form. To prepare the OH^- form remove fines and treat the resin with several additions of 1 M hydrochloric acid followed by several additions of 1 M sodium hydroxide and rinse with deionized water.

Equipment:

Use an all-glass column similar to the design of Pankow and Janauer[11] for the column experiments. Modify the system by attaching the column to a three-way stopcock by a 5/20 ground glass joint shortened to 5/12. Fill the column 1 × 6 cm with 3 ml of wet resin (the three-way stopcock allows the sample to be loaded onto the bottom of the column by ascending flow). Use descending flow to elute the metal ions through the bottom of the column. Use a Perkin-Elmer Model 360 atomic absorption spectrometer with Perkin Elmer HGA 2100 graphite furnace or equivalent.

Procedure:

The precomplexation column procedure is applicable to up to a litre sample containing up to 2×10^{-6} M heavy metal ions. For each 100 ml of sample, add 5 ml of 8-hydroxyquinoline solution and 25 ml of pH 8 Tris–HCl buffer (check final pH) and pass the solution through the column at a flow rate of 10 ml min^{-1} using diluted (1 part buffer: 4 parts water) pH 8 buffer to wash the final amounts of the sample onto the column. Elute zinc, cadmium, and lead with 2 M nitric acid; elute cobalt by adding 3 ml of 12 M hydrochloric acid followed by sufficient 2 M nitric acid to complete the elution.

This procedure gives 99 per cent recovery for zinc(II), lead(II), and cadmium(II) and a 94–99 per cent recovery for cobalt(II) at the μg $litre^{-1}$ concentration range (0.2–40 μg $litre^{-1}$). Due to the fact that the preconcentrated anion exchange columns can be stored for up to 7 days without loss of recovery on subsequent elution of metals the technique is well suited for on-site work. Samples can be collected and concentrated and the loaded resins stored for a week before elution and measurement.

Mandel and Das[12] applied an anion exchange resin to the determination of traces of mercury as an anionic complex in natural waters by cold vapour atomic absorption spectrometry using a reduction aeration method.

Reagents/materials:

Stock solution of mercury(II) chloride, 0.100 g litre^{-1} mercury prepared from BDH Analar grade material by dissolution in 5 per cent nitric acid containing 0.01 per cent potassium dichromate.

Working standards, 1.0 to 10.0 μg litre^{-1} prepared by dilution of the stock solution with 5 per cent nitric acid containing 0.01 per cent potassium dichromate.

Tin(II) chloride (E. Merck) (10 per cent w/v) freshly prepared by dissolution in hot 1:1 hydrochloric acid. Boil the solution for about 5 min to expel any mercury impurity.

Sodium chloride, analytical reagent grade previously heated to 700 °C in a muffle furnace.

Hydrochloric acid, redistilled.

Nitric acid, redistilled.

Dowex AG 1-X 8 anion exchange resin (100–200 mesh) in chloride form. Place in a glass column (1.0 cm i.d. × 25 cm long) and adjust the resin bed height to about 5 cm. Wash the column successively with 50 ml 1:1 hydrochloric acid and 50 ml 0.1 N hydrochloric acid each time before use.

Apparatus:

ECIL Model MA 5800 A mercury analyser. Connect the reduction aeration vessel to a U tube filled with anhydrous perchlorate to serve as a moisture trap, and directly connect this to the inlet tube of the instrument.

Sample preparation:

Collect water samples in polythene bottles previously cleaned by soaking overnight in 10 per cent nitric acid. Preserve the samples with 0.1 N hydrochloric acid containing 0.01 per cent potassium dichromate. Irradiate the samples with ultraviolet light for 15 min before analysis.

Preconcentration:

Transfer 5 ml of sample to a beaker and treat with hydroxylammonium chloride solution (10 per cent w/v) added drop by drop until the yellow colour due to potassium dichromate disappears. Add 50 g of sodium chloride and stir the sample to dissolve the salt. Pass the solution through the ion exchange column at a rate of 5–6 ml min^{-1}. Then wash the column with redistilled water until the eluates are acid-free. Pass 80 ml of 4 N nitric acid through the column at a rate of 3 ml min^{-1} and collect the eluate, in a 100 ml volumetric flask containing 1 ml of 1 per cent potassium dichromate solution. Make this solution up to volume with 4 N nitric acid and use for cold vapour atomic absorption measurements.

Atomic absorption measurements:

Transfer aliquots containing about 20–200 ng of mercury into the reduction aeration vessel and dilute to 50 ml with 4 N nitric acid. Add 5 ml of tin(II) chloride solution and stir the solution with a magnetic stirrer for 5 min. Purge the mercury vapour with air into the instrument for absorbance measurements.

The calibration graph obtained by this procedure was linear over the range 20–400 ng of mercury. The detection limit, expressed as twice the standard deviation of the blank value, was equivalent to 2 ng $litre^{-1}$. The coefficients of variation calculated from 10 replicate determinations of 400 and 50 ng of mercury were 2.6 and 6.8 per cent, respectively.

The method was applied to the determination of trace amounts of mercury in spring water and stream water samples. Five-litre water samples were used in each determination. Recovery of mercury from these samples was studied by addition of known amounts of mercury as mercury(II) chloride. The results are shown in Table 66.

Dowex A-1 has a serious disadvantage in that the resin also retains the alkaline earth metal ions but it has been shown experimentally

Table 66 Determination of mercury in water samples.

Sample	Sample volume (litre)	Mercury added (ng)	Mercury found (ng)[a]
Spring water A	5		47
	5	100	143
Spring water B	5		55
	5	100	152
Spring water C	5		392
	5	100	488
Stream water A	5		41
	5	100	139

[a] Mean of three replicate determinations.
From Mandel and Das[12] with permission.

that concentrations of sodium up to 5×10^{-1} M and calcium up to 4×10^{-3} M do not affect the recovery of the other trace metals.

Silver has been preconcentrated from 1-litre sample at pH 1 on an AG-1-X8 anion exchange column.[13] Silver was eluted from this column with acetone: nitric acid: water (20:1:1) and acetone removed from the elute by evaporation. After pH adjustment to 2–3 the solution was treated with ammonium pyrrolidinedithiocarbamate and the silver chelate extracted into a small volume of methyl ethyl ketone. Evaluation at 328.1 μm by atomic absorption spectrometry enabled extremely low levels of silver to be determined.

Riley and Siddiqui[14] preconcentrated thallium by adsorption of the tetrachlorothallate ion on to a strongly basic anion exchange resin, followed by elution with sulphur dioxide and determination by graphite furnace atomic absorption spectroscopy or differential pulse anodic stripping voltammetry. De Ruck *et al.*[15] oxidized thallium(I) to thallium(III) with cerium(IV) sulphate and passed the solution through an anion exchange column which retained thallium as the tetrachlorothallate(III) ion. Thallium was then eluted with ammonium sulphate solution and determined by electrothermal atomic absorption spectrometry. A concentration factor of 400 was achieved and the detection limit was 3.3 ng litre^{-1}.

Vazquez-Gonzalez *et al.*[16] have described a method for preconcentrating and determining molybdenum by electrothermal atomization atomic absorption spectrometry after preconcentration by means of anion exchange using Amberlite IRA-400 in resin citrate form. The optimal analytical parameters were established by drying, carbonization, charring, atomization, and cleaning in a graphite furnace. The precision and accuracy of the method were investigated. Less than 0.2 μg litre^{-1} molybdenum could be determined by this procedure.

Neutron activation analysis Bergerioux *et al.*[17] compared various methods including anion exchange resin techniques of preconcentration of 22 elements. In this technique the metals were adsorbed on to the resin which was then examined directly by neutron activation analysis and gamma spectrometry (when radiotracers were used). Asamov[18] used preconcentration on anion exchange resin followed by neutron activation analysis to determine total gold in natural water. Becknell *et al.*[19] adsorbed mercury from acidified natural water on to anion exchange resin loaded paper prior to analysis by neutron activation analysis. Down to 0.005 μg mercury in the original sample could be determined.

Other analytical finishes Further applications of anion exchange resins to the preconcentration of metals in non-saline waters are reviewed in Table 67.

5.2.2 Sea-water

Spectrophotometric methods Shriadah and Ohzeki[34] determined iron in sea-water by spectrophotometry after enrichment as a bathophenanthroline disulphonate complex on a thin-layer of anion exchange resin. Sea-water samples (50 ml) containing iron(II) and iron(III) were diluted to 150 ml with water followed by sequential addition of 20 per cent hydrochloric acid (100 μl), 10 per cent hydroxylammonium chloride (2 ml), 5 M ammonium solution (to pH 3.0 for iron(III) reduction), bathophenanthroline disulphonate solution (1.0 ml), and 10 per cent sodium acetate solution (2.0 ml) to give a mixture with a final pH of 4.5. A macroreticular anion exchange resin, Amberlyst A27, in the chloride form was added, the resultant coloured thin layer was scanned by a densitometer and the adsorbance measured at 550 nm.

Kodama and Tsubota[35] determined tin in sea-water by anion exchange chromatography and spectrophotometry with Catechol Violet. After adjusting to 2 mol litre^{-1} in hydrochloric acid 500 ml of the sample is adsorbed on a column of Dowex 1-Xs resin (Cl-form) and elution is then effected with 2 M nitric acid. The solution is evaporated to dryness after adding 1 M hydrochloric acid and the tin is again adsorbed on the same column. Tin is eluted with 2 M nitric acid. Tin is determined in the eluate by the spectrophotometric Catechol Violet method. There is no interference from 0.1 mg of each of aluminium, manganese, nickel, copper, zinc, arsenic, cadmium, bismuth, and uranium, any titanium, zirconium, and antimony are removed by the ion exchange. Filtration of the sample through a Millipore filter does not affect the results, which are in agreement with those obtained by neutron activation analysis.

Kiriyama and Kuroda[36] combined ion exchange preconcentration with spectrophotometry using 2-pyridylazoresorcinol in the determination of vanadium in sea-water. The sample (2 litres) made 0.1 M in hydrochloric acid, filtered, and made 0.1 M in ammonium thiocyanate, is passed through a column of Dowex 1-X8 resin anion exchange (thiocyanate form). The vanadium is retained and is eluted with concentrated hydrochloric acid. Thiocyanate in the elute is decomposed by heating with nitric acid and the solution is evaporated to fuming with sulphuric acid. A solution of the residue is neutralized with aqueous ammonia and evaporated nearly to dryness. The residue is treated with water and aqueous sodium hypobromite and after 30 min with phenol, phosphate buffer solution of pH 6.5, and aqueous 1.2-diaminocyclohexane-*N*, *N*, *N'* *N'*-tetra-acetic acid, and the vanadium is determined spectrophotometrically at 545 nm with 4-(2-pyridylazo) resorcinol. Vanadium was determined in sea-water at levels of 1.65 μg litre^{-1}. After boiling such samples under reflux with potassium per-

Table 67 Preconcentration of anionic metal species from natural water.

Metal	Type of water	Resin	Medium	Eluting agent	Detection limit	Analytical finish	Ref.
Al	Potable	Anion exchange	Pyrocatechol violet	–	–	–	20
Fe	Natural	Anion	–	–	1–2 ng	Spectrophotometric	21
Cd	Natural	Amberlite IRA-400	Cyanide	Acetic acid	–	Spectrophotometric	22
Zn	Sewage	EDE-10 P	Hydrochloric acid	Hydrochloric acid	–	Spectrophotometric	23
Co	Natural	Dowex AG1-X8	Hydrochloric acid–thiocyanate	Hydrochloric acid	–	Spectrophotometric	24
Cd	Natural	Dowex AG1-X8	Hydrochloric or hydrobromic acid	Nitric acid	–	Spectrophotometric	25

Th	Natural	Dowex AG1-X8	Nitric acid	Hydrochloric acid	–	Spectrophotometric	26
Ni	Mineral water	Dowex A-1	Sodium acetate	Hydrochloric acid	0.5 μg	Spectrophotometric	27
U	Natural	Dowex AG1-X8	Sulphuric acid	Sulphuric acid	0.3 μg $litre^{-1}$	Fluorimetric	28
U	Natural	Amberlite LA-1	–	–	0.04 μg $litre^{-1}$	Fluorimetric	29–31
^{239}Pu ^{240}Pu		Anion exchange	–	Hydrofluoric acid–hydrochloric acid	–	Deposited on Pt–γ-ray spectrometry	32
Various radio nucleids		Single bead of anion exchange resin	–	–	–	Point source mass spectrometry	33

manganate and sulphuric acid (to establish the concentration of organically bound vanadium), values for vanadium were 30–60 per cent higher than corresponding values obtained without oxidation.

Anion exchange resins have been used to preconcentrate molybdenum is sea-water prior to its spectrophotometric determination as the Tiron complex.[37-39] Kawabuchi and Kuroda[40] have concentrated molybdenum by anion exchange from sea-water containing acid and thiocyanate or hydrogen peroxide[40,41] and determined it spectrophotometrically. Korkisch *et al.*[42] have concentrated molybdenum from natural waters on Dowex 1-X8 anion exchange resin in the presence of thiocyanate and ascorbic acid. A sodium citrate and ascorbic acid system has also been worked out for the concentration of molybdenum on Dowex 1-X8 (citrate form) as a citrate complex from tap and mineral waters.

In method described by Kiriyama and Kuroda[43] molybdenum is sorbed strongly on Amberlite CG 400 anion exchange resin (Cl form) at pH 3 from sea-water containing ascorbic acid and is easily eluted with 6 M nitric acid. Molybdenum in the effluent can be determined spectrophotometrically with potassium thiocyanate and stannous chloride. The combined method allows selective and sensitive determination of traces of molybdenum in sea-water. The precision of the method is 2 per cent at a molybdenum level of 10 μg litre^{-1}. To evaluate the feasibility of this method Kiriyama and Kuroda[43] spiked a known amount of molybdenum into sea-water and analysed it by the procedure, the results are given in Table 68. As can be seen, the recoveries for the molybdenum added to 500 or 1000 ml samples are satisfactory.

Shriadah *et al.*[44] determined molybdenum(VI) in sea-water spectrophotometrically after enrichment as the Tiron complex on a thin layer of anion exchange resin. There were no interferences from trace elements or major constituents of sea-water, except for chromium and vanadium. These were reduced by the addition of ascorbic acid. The concentration of dissolved molybdenum(VI) determined in Japanese sea-water was 11.5 μg litre^{-1} with a relative standard deviation of 1.1 per cent.

Kuroda *et al.*[45] observed that uranium is strong adsorbed from acidified saline solutions by a strongly basic ion exchange resin in the presence of azide ions. The distribution coefficient of uranium with 0.5 M sodium chloride increased rapidly with an increase in azide concentration, and was much higher than the coefficient obtained with hydrazoic acid alone. The sorbed uranium was easily eluted with 1 M hydrochloric acid, and was determined spectrophotometrically. Recoveries of 98.3–99.6 μg uranium per litre were obtained from artificial sea-water containing 3.4 μg uranium per litre. Selenium(IV) has been preconcentrated on a bismuthiol(II) modified anion exchange resin (Amberlite IRA-400)[46] followed by fluorometric estimation using diamino napthalene. Selenium(IV) adsorbed on the column as selentotri-

Table 68 Determination of molybdenum in sea-water.

Sample[a]	Sample volume (litres)	Mo added (μg)	Mo found (μg)	Original content (μg litre^{-1})
Sea-water A	0.5	0	4.27, 4.51	8.54, 9.02
			4.39	8.78
	1.0	0	8.90	8.90
	0.5	4.24	8.73	8.98
	0.5	8.48	12.80	8.64
				av. 8.81 ± 0.19
Sea-water B	0.5	0	4.85, 4.70	9.70, 9.40
			4.85	9.70
	1.0	0	9.48	9.48
	0.5	4.24	9.02	9.56
	0.5	8.48	13.20	9.44
				av. 9.55 ± 0.13
Sea-water C	0.5	0	4.11, 4.08	8.22, 8.16
			4.29	8.58
	1.0	0	8.54	8.54
	0.5	4.24	8.44	8.40
	0.5	8.48	12.70	8.44
				av. 8.39 ± 0.17

[a] Sea-water A: collected at Kamoike Harbour, Kagoshima Bay, Japan, on 23 June 1983, Slinity 33.48 per cent.
Sea-water B: collected on the shore at Kushikino, East China Sea, on 30 June 1983, Salinity 34.03 per cent.
Sea-water C: collected at Yamagawa Harbour, Kagoshima Bay, on 6 July 1983, Salinity 31.55 per cent.
From Kiriyama and Kuroda[43] with permission.

sulphate was desorbed with a small volume of 0.1 M penicillamine or 0.1 M cysteine prior to fluorimetric determination of selenium.[33]

Atomic absorption spectrometry A limited amount of work has been carried out using polyacrylamidoxine resin. Collela *et al.*[47] agitated sea-water samples with polyacrylamidoxime resin preparatory to the determination of iron(III), copper(II), cadmium(II), lead(II), and zinc(II). Metals were removed from the filtered OH resin by equilibrating with 1:1 hydrochloric acid/water mixture and their concentrations determined by atomic absorption spectrometry. Metal concentrations as determined by the resin method were in good agreement with the values determined directly on samples by either differential pulse polarography or differential pulse anodic stripping voltammetry.

Kiriyama and Kuroda[48] determined vanadium, cobalt, copper, zinc, and cadmium in sea-water by adsorption in an anion exchange resin. Preconcentration from sea-water was achieved in thiocyanate medium. A strongly basic anion exchange resin in the thiocyanate from concentrated the five metals from sea-water adjusted to 1 M thiocyanate and 0.1 M hydrochloric acid. Sorbed metals were recovered simultaneously by elution with 2 M nitric acid prior to determination by graphite furnace atomic absorption spectrometry.

Adsorption of metals on a single bead of ion-exchange resin has been used as a means of effecting preconcentration.[49,50] Koide *et al.*[49] showed that cadmium, palladium, iridium, gold, plutonium, and technetium can be concentrated from sea-water on to a single bead of anion exchange resin. This process eliminated salt interference. The beads acted as point sources during subsequent analytical determination. Optimal conditions for the adsorption of cadmium-109, palladium-103, iridium-192, gold-105, plutonium-237, and technetium-95m on to a single bead were determined. Two types of applications of the techniques were investigated, with no prior concentration and with preconcentration; increasing the yield of plutonium and technetium on to a single bead for improved sensitivity in mass spectrometric analysis. Two types of anion exchange resin (gel type and macroporous type) were tested. Hodge *et al.*[50] determined platinum and iridium in marine waters by preconcentration by anion exchange, purification by uptake on a single anion exchange bead, and determination by graphite furnace atomic absorption spectrometry. All steps were followed by radiotracers (platinum-191 and iridium-192). Yields varied between 35 and 90 per cent for determination of platinum and iridium in sediments, manganese nodules, sea-water, and microalgae.

Kuroda *et al.*[51] preconcentrated trace amounts of molybdenum from acidified sea-water on a strongly basic anion exchange resin (Bio-Rad AGI-X8 in the chloride form) by treating the water with sodium azide. Molybdenum(VI) complexes with azide were stripped from the resin by elution with ammonium chloride/ammonium hydroxide solution (2 M to 2 M). Relative standard deviations of better than 8 per cent at levels of 10 μg litre^{-1} were attained for sea-water using graphite furnace atomic absorption spectrometry.

Chelation solvent extraction or resin-adsorption preconcentration procedures coupled with graphite furnace atomic absorption spectrometry or neutron activation analysis are capable of determining many elements at the baseline concentrations occurring in deep sea-water samples. All workers in this field emphasize the need for extreme precautions during sampling of sea-water for analysis at these very low concentrations.

Two methods for the determination of vanadium in sea-water have been studied which use neutron activation analysis and atomic absorption spectrometry.[52] In the atomic absorption spectrometric procedure,[52] potassium thiocyanate (10 g) and ascorbic acid (5 g to reduce to V^{VI}) were dissolved in 1 litre of sea-water and the solutions were left to stand for 2–3 h. These samples (1–3 litres) were passed through a Dionex 1-X8 anion exchange column at a flow rate of 1.7 ml min^{-1}. The resin was then washed with 20 ml distilled water and vanadium eluted with 150 ml eluent solution. The vanadium eluate was slowly evaporated under an infra-red lamp, the residue dissolved in 10 ml 6 M hydrochloric acid containing 1 ml of the aluminium chloride solution[53] and vanadium determined by atomic absorption spectrophotometry. For calibration, suitable standard solutions were aspirated before and after each batch of samples.

The analysis of the sea-water samples by both methods is shown in Table 69. The average concentration and standard deviation of the Pacific Ocean waters (μg $litre^{-1}$) were 2.00 ± 0.09 by neutron activation analysis and 1.86 ± 0.12 by atomic absorption spectrometry. For the Adriatic water the corresponding values were about 1.7 μg $litre^{-1}$. The difference between the values for the same sea-water is within the range to be expected from the standard deviations observed. Though the neutron activation analysis is inherently more sensitive than the atomic absorption spectrometry, both procedures give a reliable measurement of vanadium in sea-water at the natural levels of concentration.

Table 69 Results of vanadium determinations in sea-water samples.

Sample	Volume (litres)	Vanadium concentration (μg $litre^{-1}$)	
		NAA	AAS
Pacific Ocean (Scripps Pier)	1		1.80
	1		2.00[a] (2.0)
	2		1.90[a]
	3		1.73
	0.098	1.99[b]	
	0.098	2.00[c] (0.20)	
Adriatic Sea			
(Shore near Lignano	3		1.71
Sabbiadoro, Italy)	0.041	1.69	
(Shore near Ancona, Italy)	3		1.73
	0.043	1.64	

[a] Results after subtraction of the quantity of vanadium added to the sample before the ion-exchange or co-precipitation step. The amount added (in μg) is shown in parentheses.
[b] Average of 12 determinations, standard deviation 0.10 μg $litre^{-1}$.
[c] Average of two samples, average deviation 0.01 μg $litre^{-1}$.
From Weis *et al.*[52] with permission.

Neutron activation analysis Matthews and Riley[54] have described the following procedure for determining down to 0.06 μg litre^{-1} rhenium in sea-water. From 6 to 8 μg litre^{-1} rhenium was found in Atlantic sea-water. The rhenium in a 15-litre sample of sea-water acidified with hydrochloric acid, is concentrated by adsorption on a column of Deacidite FF anion exchange resin (Cl-form), followed by elution with 4 M nitric acid and evaporation of the eluate. The residue (0.2 ml) together with standards and blanks, is irradiated in a thermal neutron flux of at least 3×10^{12} neutrons cm^{-2} s^{-1} for at least 50 h. After a decay period of 2 days, the sample solution and blank are treated with potassium perrhenate as carrier and evaporated to dryness with a slight excess of sodium hydroxide. Each residue is dissolved in 5 M sodium hydroxide. Hydroxylammonium chloride is added (to reduce Tc^{VII}) which arises as ^{99m}Tc from activation of molybdenum present in the samples, and the Re^{VII} is extracted selectively with ethyl methyl ketone. The extracts are evaporated, the residue is dissolved in formic acid–hydrochloric acid (19:1), the rhenium is adsorbed on a column of Dowex 1, and the column is washed with the same acid mixture followed by water and 0.5 M hydrochloric acid; the rhenium is eluted at 0 °C with acetone–hydrochloric acid (19:1) and is finally isolated by precipitation as tetraphenylarsonium perrhenate. The precipitate is weighed to determine the chemical yield and the ^{186}Re activity is counted with an end-window Geiger-Müller tube. The irradiated standards are dissolved in water together with potassium perrhenate. At a level of 0.057 μg litre^{-1} rhenium the coefficient of variation was $\pm$7 per cent.

Kawabuchi and Riley[55] used neutron activation analysis to determine silver in sea-water. Silver in a 4 litre sample of sea-water was concentrated by ion exchange on a column (6 cm $\times$ 0.8 cm) containing 2 g of Deacidite FF-IP anion exchange resin, previously treated with 50 ml 0.1 M hydrochloric acid. The silver was eluted with 20 ml 0.4 M aqueous thiourea and the eluate was evaporated to dryness, transferred to a silica irradiation capsule, heated at 200 °C, and ashed at 500 °C. After sealing, the capsule was irradiated for 24 h in a thermal-neutron flux of 3.5×10^{12} neutrons cm^{-2} s^{-1}, and after a decay period of 2–3 days, the ^{110m}Ag arising from the reaction ^{199m}Ag(n, γ) ^{110m}Ag was separated by a conventional radiochemical procedure. The activity of the ^{110m}Ag was counted with an end-window Geiger-Müller tube, and the purity of the final precipitate was checked with a Ge(Li) detector coupled to a 400-channel analyser. The method gave a coefficient of variation of $\pm$ 10 per cent at a level of 40 ng silver per litre.

5.3 Organics – non-saline waters

Lee *et al.*[56] preconcentrated phenols and amino phenols from industrial waste waters using Dowex 1-X8 anion exchange resin. Gardner and Lee[57] preconcentrated free and combined amino acids from lake water on an ion exchange column, prior to desorption with a small volume of acid, conversion to *N*-trifluoroacetyl methyl esters, and gas chromatography.

Nitriloacetic acid has been preconcentrated from water at pH 3 by passage through an anion exchange column and subsequent elution with sodium chloride[58] prior to estimation by polarography. Longbottom[59] preconcentrated niltriloacetnic acid on a strong anion exchange resin, then desorbed it with sodium tetraborate at pH 9, prior to estimation by ultraviolet spectroscopy. Weber and Wilson[60] used anion and cation exchange resins to preconcentrate fulvic and humic acids from water.

Insecticides containing PH_3 after acid hydrolyses to free methylphosphoric acids have been preconcentrated on anion exchange resins prior to desorption, conversion to their dimethyl esters and estimation by gas chromatography.[61] In a UK standard method for preconcentrating diquat and paraquat in river and potable waters the herbicides and preconcentrated on an anion exchange resin, then reduced with sodium dithionite prior to measurement at 374 nm by visible spectrophotometry.[62]

Condo and Janauer[63] give details of a method of preconcentrating and determining traces of Aldicarb in water. The method involves decomposition of aldicarb by passage through a strongly basic anion exchange resin to produce an oximate which was adsorbed on the resin. Subsequent addition of sulphuric acid resulted in in-situ formation and elution of hydroxylamine, which was used for quantitative reduction of trivalent iron to the ferrous state; this was then determined spectrophotometrically.

5.4 Organometallic compounds

Organomercury compounds have been preconcentrated by conversion to tetrachloromercury derivative $HgCl_4^{2-}$ and passage through a filter disc loaded with SB-2 anion exchange resin.[64] Mercury was then determined in the resin by neutron activation analysis. Ahmed *et al.*[65] preconcentrated methylmercury and inorganic mercury from rain water on an anion exchange column prior to analysis by cold vapour atomic absorption spectrometry. Methylmercury was determined by passing a 500 ml sample (in 5 per cent hydrochloric acid) through an anion exchange column on which ionic mercury was retained. Methylmercury passing the column was decomposed by UV and the methylmercury concentration determined by difference. Recovery of ionic mercury increased as acid concentration increased and with UV irradiation.

References

1. Parkow, J. F. and Janauer, G. E. (1974). *Analytica Chimica Acta*, *69*, 97.
2. Mazzucotelli, A., Minoia, C., Pozzoli, L., and Ariati, L. (1982). *Applied Spectroscopy*, *4*, 182.
3. Parkow, J. F., Lieta, D. P., Lin, J. W., Ohl, S. E., Shum, W. P., and Janauer, G. E. (1977). *Science of the Total Environment*, *7*, 17.
4. Dankow, J. F. and Janauer, G. E., (1974). *Analytica Chimica Acta*, *69*, 97.
5. Westland, A. D. and Boiichair, I. (1974). *Water Research*, *8*, 467.
6. Blancher, R. W. and Riego, D. (1975). *Journal of Environmental Quality*, *4*, 45.
7. Paez, D. M. and Guagnini, O. A. (1971). *Mikrochimica Acta*, *2*, 220.
8. Jun, Z., Oshima, M., and Motomizu, S. (1988). *Analyst (London)*, *113*, 1631.
9. Korkische, J. and Sario, A. (1977). *Analytica Chimica Acta*, *76*, 393.
10. Berge, D. G. and Going, J. E. (1981). *Analytica Chimica Acta*, *123*, 19.
11. Pankow, J. F. and Janauer, J. E. (1974). *Analytica Chimica Acta*, *69*, 97.
12. Mandel, S. and Das, A. K. (1982). *Atomic Spectroscopy*, *3*, 56.
13. Chau, I. T., Fishmann, M. J., and Ball, T. W. (1969). *Analytica Chimica Acta*, *43*, 189.
14. Riley, J. P. and Siddiqui, S. A. (1986). *Analytica Chimica Acta*, *181*, 177.
15. De Ruck, A., Vandecasteele, C., and Dams, R. (1987). *Mikrochimica Acta*, No. 416, 187.
16. Vazquez-Gonzalez, J. F., Bermejo-Barrera, P., and Bermejo-Martinez, F. (1987). *Atomic Spectroscopy*, *8*, 159.
17. Bergerioux, C., Blan, P. C., and Haerdi, W. (1977). *Journal of Radioanalytical Chemistry*, *39*, 823.
18. Asamov, K. A., Abdullah, A. A., and Zakhidov, A. S. (1969). *Doklady Acad. Nauk. Uzbek. SSR*, (3), 26.
19. Becknell, D. E., Marsh, R. H., and Allie, W. (1971). *Analytical Chemistry*, *43*, 1230.
20. Sarzanini, C., Mentasti, E., Porta, V., and Gennaro, M. C. (1987). *Analytical Chemistry*, *59*, 484.
21. Shi Yu, L. and Wei Ping, G. (1984). *Talanta*, *31*, 844.
22. Ashizawa, T. and Hosoya, K. (1971). *Japan Analyst*, *20*, 1416.
23. Kurochkina, M. I., Lyakh, V. I., and Perelyaeva, G. L. (1972). *Nauch. Trudy. irkutsh. gos. Nauchno issled. Inst. Tedk. Esvet. Metall*, (24), 149, Ref: *Zhur. Khim.* 19 GD (13) (1972) Abstract No 13 G 148.
24. Korkische, J. and Dimitriades, D. (1973). *Talanta*, *20*, 1287.
25. Korkische, J. and Dimitriades, D. (1973). *Talanta, 20*, 1295.

26. Korkisde, J. and Dimitriades, D. (1973). *Talanta*, *20*, 1303.
27. Nevoral, V. and Okae, A. (1988). *Czlka Form*, *17*, 478.
28. Danielsson, A., Roennholm, B., Kiellstoem, L. E., and Ingman, F. (1973). *Talanta*, *20*, 185.
29. Brits, R. J. S. and Smit, M. C. B. (1977). *Analytical Chemistry*, *49*, 67.
30. Gladney, E. S., Owens, J. W., and Starner, J. W. (1976). *Analytical Chemistry*, *48*, 973.
31. Zielinski, R. A. and McKoun, M. (1984). *Journal of Radioanalytical and Nuclear Chemistry Articles*, *84*, 207.
32. Golchert, N. V. and Sedlet, J. (1972). *Radiochemical and Radio analytical Letters*, *12*, 215.
33. Carter, J. A., Walker, R. L., Smith, D. H., and Christie, W. H. (1980). *International Journal of Environmental Analytical Chemistry*, *8*, 241.
34. Shriadah, M. M. A. and Ohzeki, K. (1986). *Analyst (London)*, *111*, 555.
35. Kodama, Y. and Tsubota, H. (1971). *Japan Analyst*, *20*, 1554.
36. Kiriyama, T. and Kuroda, R. (1972). *Analytica Chimica Acta*, *62*, 464.
37. Shriadah, H. M. A., Katoeka, M., and Ohzeri, K. (1985). *Analyst (London)*, *110*, 125.
38. Riley, J. P. and Taylor, D. (1968). *Analytica Chimica Acta*, *40*, 479.
39. Kawabuchi, K. and Kuroda, K. (1969). *Analytica Chimica Acta*, *46*, 23.
40. Kawabuchi, K. and Kuroda, R. (1969). *Analytica Chimica Acta*, *46*, 23.
41. Kuroda, R. and Tarui, T. (1974). *Fresenius Zeitschrift fur Analytische Chemie*, *269*, 22.
42. Korkische, J., Godl, L., and Gross, H. (1975). *Talanta*, *22*, 669.
43. Kiriyama, R. and Kuroda, R. (1984). *Talanta*, *31*, 472.
44. Shriadah, H. M. A., Katoaka, M., and Ohzeki, K. (1985). *Analyst (London)*, *110*, 125.
45. Kuroda, R., Oguma, K., Mukai, N., and Imamoto, M. (1987). *Talanta*, *34*, 433.
46. Wu, T. L., Lambert, L., Hastings, D., and Banning, D. (1980). *Bulletin Environmental Contamination and Toxicology*, *24*, 411.
47. Colella, M. B., Siggia, S., and Barnes, R. m. (1980). *Analytical Chemistry*, *52*, 2347.
48. Kiriyama, T. and Kuroda, R. (1985). *Mikrochimica Acta*, *1*, 405.
49. Koide, M., Lee, D. S., and Stallard, M. O. (1984). *Analytical Chemistry*, *56*, 1956.
50. Hodge, V., Stallard, M., Koide, M., and Goldberg, E. D. (1986). *Analytical Chemistry*, *58*, 616.
51. Kuroda, R., Matsumoto, N., and Ogmura, K. (1988). *Fresenius Zeitschrift fur Analytische Chemie*, *330*, 111.
52. Weiss, H. V., Güttman, H. A., Korkische, J., and Steffan, I. (1977). *Talanta*, *24*, 509.
53. Korkische, J. and Gross, H. (1973). *Talanta*, *20*, 1153.
54. Matthews, A. D. and Riley, J. P. (1970). *Analytica Chimica Acta*, *51*, 455.
55. Kawabuchi, K. and Riley, J. P. (1973). *Analytica Chimica Acta*, *65*, 271.
56. Lee, K. S., Lee, D. W., and Chung, Y. S. (1973). *Analytical Chemistry*, *45*, 396.

57. Gardner, W. S. and Lee, G. E. (1973). *Journal Environmental Science and Technology*, *7*, 719.
58. Haberman, J. P. (1971). *Analytical Chemistry*, *43*, 63.
59. Longbottom, J. E. (1972). *Analytical Chemistry*, *44*, 418.
60. Weber, J. H. and Wilson, S. A. (1975). *Water Research*, *9*, 1079.
61. Verweij, J. A., Regenhardt, C. E. A., and Boter, H. L. (1970). *Chemosphere*, *8*, 115.
62. Her Majesty's Stationery Office, Methods for the Examination of Waters and Associated Materials. 1988. (1987). In *Determination of Diquat and Paraquat in River and Drinking Waters, spectrophotometric methods tentative*, HMSO, London.
63. Condo, D. P. and Janauer, G. E. (1987). *Analyst (London)*, *112*, 1027.
64. Becknell, D. E., Marsh, R. H., and Allie, W. (1971). *Analytical Chemistry*, *43*, 1230.
65. Ahmed, R., May, K., and Stoeppler, M. (1987). *Fresenius Zeitschrift fur Analytische Chemie*, *326*, 510.

INDEX